Y0-BYK-473

TEACHING GENETICS

in School and University

TEACHING GENETICS

in School and University

EDITED BY

C. D. DARLINGTON

AND A. D. BRADSHAW

OLIVER & BOYD

EDINBURGH AND LONDON

OLIVER AND BOYD LTD

Tweeddale Court
Edinburgh 1

39a Welbeck Street
London W.1

First published 1963

PRINTED IN GREAT BRITAIN BY
HAZELL WATSON AND VINEY LTD
AYLESBURY, BUCKS

INTRODUCTION

The teaching of any scientific subject should have two aims: to explain its foundations and to show how it is being advanced.

Genetics is a young science: the term was introduced by Bateson in 1906. The foundations and the purpose of the subject remain the same as when it began. Anyone who teaches Mendelism and chromosome behaviour cannot yet be accused of being reactionary or academic.

But the youth of genetics brings other problems. The development is as rapid as ever: the last two decades have seen the development of microbial, biochemical and population genetics, the analysis of breeding systems and of continuous variation and the understanding of chromosome structure, activity and breakage. These new discoveries have amplified the ideas and broadened the scope of the subject: it now touches biology at every point, plant and animal, human and microbial, physiological, ecological and evolutionary.

For these reasons a number of geneticists working in universities in this country suggested that a symposium on teaching methods would be useful. It would bring together those concerned with particular aspects of the subject and those who had to teach the whole subject. There would be a chance to discover the best modes and aims of attack and the best materials to use for experiment and demonstration. This would make it possible to teach more accurately, more vividly and, not least important, more easily.

The symposium aroused great interest and we are therefore publishing the contributions. At the same time, for the use of a wider public, notes have been added on genetics in schools and on certain demonstrations.

Finally the account has been completed with a list of sources of materials mainly compiled by Dr. Ralph Riley, Secretary of the British Genetical Society.

C. D. Darlington
A. D. Bradshaw

CONTRIBUTORS

BEVAN, E. A. . . . . Botany School, Oxford.

BRADSHAW, A. D. . . . . . . University College of North Wales, Bangor.

CLOWES, R. C. . . . . M.R.C. Microbial Genetics Unit, Hammersmith Hospital, London, W.12

CROWE, L. K. . . . . Botany School, Oxford.

DARLINGTON, C. D. . . . . Botany School, Oxford.

DAVIES, W. ELLIS . . . . . . Welsh Plant Breeding Station, Aberystwyth.

FALCONER, D. S. . . . . . . University College of North Wales, Bangor.

JINKS, J. L. . . . . A.R.C. Unit of Biometrical Genetics, Birmingham University.

JOHN, B. . . . . Department of Genetics, Birmingham University.

LEWIS, K. R. . . . . Botany School, Oxford.

MOURANT, A. E. . . . . M.C.R. Blood Group Laboratory, Lister Institute, London, S.W.1.

OCKEY, C. H. . . . . Christie Hospital and Holt Radium Institute, Withington, Manchester.

PATEMAN, J. A. . . . . Department of Genetics, Milton Road, Cambridge.

PUSEY, J. G. . . . . Botany School, Oxford.

REES, H. . . . . Department of Agricultural Botany, University College of Wales, Aberystwyth.

THODAY, J. M. . . . . Department of Genetics, Milton Road, Cambridge.

WHITEHOUSE, H. L. K. . . . Botany School, Cambridge.

WHITTINGTON, W. J. . . . School of Agriculture, Sutton Bonington,
University of Nottingham.

WOODS, R. A. . . . . Department of Genetics,
University of Sheffield.

WYLIE, ANN P. . . . . University of Otago, Dunedin,
New Zealand.

CONTENTS

I. GENERAL PAPERS

Introduction v
List of Contributors vii
Genetics Teaching in Universities. *J. M. Thoday* 1
Bacteria and Bacteriophages. *R. C. Clowes* 7
Transduction in Bacteria. *E. A. Bevan* 17
Fungal Genetics. *J. A. Pateman* 22
Yeast in Practical Genetics. *E. A. Bevan and R. A. Woods* 29
Biometrical Genetics. *J. L. Jinks* 36
A Practical Exercise in Quantitative Genetics. *J. M. Thoday* 41
The Use of Mice in Teaching Genetics. *D. S. Falconer* 44
Teaching Cytology. *H. Rees* 50
Demonstrating Chromosome Behaviour. *K. R. Lewis and B. John* 56
Material for Practical Cytology. *Ann P. Wylie* 59
Peripheral Blood Cultures of Human Chromosomes. *C. H. Ockey* 64
Blood Groups. *A. E. Mourant* 68
The Genetic Garden. *C. D. Darlington* 74
Genetics in Schools. *A. D. Bradshaw* 79

II. BRIEF NOTES

Seedling Characters. *Leslie K. Crowe* 90
Leaf Markings in *Trifolium repens*. *W. Ellis Davies* 94
Cyanogenesis in *Trifolium repens*. *J. G. Pusey* 99
Three Teaching Projects. *A. D. Bradshaw* 105
A *Drosophila* Population Cage. *W. J. Whittington* 110
A Practical Examination Model. *H. L. K. Whitehouse* 112

III. SOURCES OF MATERIALS

IV. CHROMOSOME FILMS

V. BOOKS ON THE TEACHING OF GENETICS

LIST OF PLATES

	facing page
PLATE 1. Chromosomes of *Paeonia* and *Allium*.	57
PLATE 2. Human Chromosomes	67
PLATES 3 and 4. Leaf markings in white clover	96–97

THE ORGANISATION OF GENETICS TEACHING IN UNIVERSITIES

J. M. THODAY
Department of Genetics
Milton Road, Cambridge

It seems likely that the spread of genetics in British Universities that has occurred since the war is gathering momentum and hence that new opportunities will arise in the next few years for the establishment and development of genetical units in a number of places. It is therefore appropriate not only that we should at this time have some discussion of the detail of genetics teaching, but that we should also give some thought to the role of genetics in University curricula and how the teaching of genetics may best be organised.

The problem is not simple, for genetics is a peculiar subject in that it concerns itself with part of the subject matter but not all the subject matter of many other subjects. It has something of value to say in fields conventionally the responsibility of departments of Zoology, Botany, Geology, Psychology, Sociology, Medicine, Biochemistry, Physiology, Anthropology and even Philology, and yet it is at the same time a remarkably coherent whole. It is therefore not surprising that attitudes to genetics teaching vary greatly, both among geneticists, and among non-geneticists. At one end of the range of attitudes is the extreme view, held by some geneticists, that all biology is genetics. (Perhaps they mean all worth-while biology, but I do not see how they judge what will be worth while 20 years hence or how they would deal with the vocational training of such as parasitologists or physiologists.) At the other end of the range is the equally extreme (but dying) view of some biologists that genetics has so little contact with what they regard as real botany or zoology that genetics can safely be left out of botanical or zoological training. Such an attitude is more common in some of the other disciplines with which genetics should have contact but as yet has little, at least on the educational side.

Neither of these views makes sense. Genetics is relevant to most biological problems, but it is patently false to equate it with biology. Genetics is both smaller and larger than biology, as biology is handled in Universities today, and what we need is for biology teaching to be directed towards educating all biologists of whatever speciality to see the truth that every biological problem can be approached from many sides, that the solution of that problem requires mutually informed approaches from all possible sides, and that at least one of the sides of each problem is genetical. Furthermore, from our point of view, some of the problems on which we can throw light, are not biological problems in the ordinary sense: for example, the genetic components of language and the relation between social structure and possible genetic differences in psychological characteristics of social classes are matters of importance we must not neglect in our consideration of the way genetics teaching should be organised. Teaching plans must look to the future rather than the past and present: we are educating the next generation.

It is this combination of the unity of genetics as a discipline and its relevance to many other subjects which presents the problem in organising genetics teaching. How are we to solve it?

From the purely administrative point of view, genetics can be provided for in two alternative ways. Either a Genetics department may be set up, or each of the departments to which genetics is relevant has its own geneticist.

The system in which each department has its tame geneticist is unsatisfactory both from the geneticists' point of view, and from the point of view of University teaching. From the geneticists' point of view it is unsatisfactory for it deprives genetics of its coherence, and prevents the genetical units becoming large enough to come alive as research units with adequate facilities and adequate numbers for the workers to stimulate one another. From the University point of view it is unsatisfactory partly because it is uneconomical of teaching time, each department having to teach its own Mendelism and chromosome theory, many students hearing the same stuff in slightly different contexts from two or three teachers. It is also unsatisfactory from the general educationists' point of view for it makes the geneticists specialists and does not allow for one of the most valuable contributions genetics can make in education by bridging gaps such as that dividing the treatment of plants and animals. The broad principles of genetics are of very general application, and to talk of plant genetics and animal genetics is to cut right across such natural divisions of genetics as there are. Such divisions have, of course, been imposed upon genetics but they have generally if not always been to the detriment of the work involved unless the workers themselves have ignored them. The place for geneticists is not in a Botany or a Zoology department. A Genetics department is required for both teaching and research.

A Genetics department, however, should not be viewed as just another department to be organised on the same basis as the established departments, for it is in a position, more than others, of overlapping their subject matter. This becomes clear when we consider who the department should teach. There are two possibilities. The department might ask for genetics to be a subject of the faculties and have its own elementary courses and special honours courses. This I wish to suggest will not prove the most satisfactory solution, either for the department or the University. From the departmental point of view it will only recruit to its elementary courses a few who start their University career with a desire to do genetics and a few who cannot think what other subsidiary course to do. In the nature of University arrangement most students will continue to do traditional combinations of subjects because they are forced with a choice of genetics *or* some other subject. From the University point of view such a system will not fulfil the function the department could fulfil of providing biologists, medicals, biochemists, sociologists and so on with a concept of what genetics can do for them in their subjects.

The alternative arrangement is the arrangement that has proved successful in Birmingham and in Sheffield and I believe in Glasgow. This is for the department of Genetics to have no elementary students 'of its own', but to act as a service department providing appropriately conceived genetics courses which are integrated into the time-tables of the main teaching departments. Thus in Sheffield, biologists (zoology and botany students taken together) have elementary courses in general genetics which are an integral part of the Botany and Zoology timetables and in the final year a corresponding joint advanced course for special honours students. Comparable courses have also been provided for the Medical School, the school of Biochemistry, and sometimes for sociologists, for psychologists and students studying for the Diploma of Public Health. Some of these courses were short, but they could be tailored to the special interests of the students concerned, and the system had the advantage that all biologists, all medicals, all biochemists, all psychologists and so on had to do some genetics and the genetics was given by people who were really interested in genetics. The consequence was establishment and growth of the Genetics department, a reasonable supply of research students, and a great deal of interest in various parts of the University in the genetical attitude to their subjects, so that it could be said that the Genetics department was serving a very important function in the University by helping to break down the barriers between subjects. This is something greatly needed in Universities today and something which genetics is peculiarly well suited to do.

This means, and this is the core of what I am trying to say, that the first duty of a University teacher in genetics must be to teach and interest people who are never going to be geneticists.

Now to some it may seem rather uninspiring and limited to set oneself as one's main task the teaching of genetics to those who are never going to be geneticists. But it is not so. In the first place it does not work like that, for if one but sits down and asks, how am I going to show these plant physiologists, these medicals or these sociologists that this and that aspect of genetics is interesting from their point of view, some of them get interested in and become geneticists. In addition there are indirect and more important rewards. For if such teaching is adequately done, sooner or later some students begin to think about the conflicting or complementary attitudes of the genetical and (say) the medical teacher to their problems. Thus are hybrid disciplines born, though the hybridisation is more likely to be introgressive. This is one of the real rewards of the educator.

A Genetics department functioning like this will be a co-operative department, co-operating in the teaching of other departments. Similar co-operation in the teaching of genetics was called for long ago (Ashby *et al.*, 1936). Its success will be dependent upon it being staffed accordingly, and its staffing must be done with a view not only to its research projects, but with a view to the teaching needs other departments have or should have. It can be taken as axiomatic that in most Universities a department wholly concerned with research would not be practicable even if it were desirable.

Now whether or not they want it, those disciplines that most need an element of teaching in genetics are Botany, Zoology, Biochemistry, Geology, Psychology, Sociology, Medicine and Agriculture. The major divisions of genetics are broadly the biochemical, the developmental, the evolutionary or population aspects and the cytological, and though biochemists will tend to view biochemical genetics as the only important aspect, to all the other departments concerned, population genetics in the broadest sense and developmental genetics are more essential. Neither of these can, of course, be adequately covered without some consideration of biochemical genetics and cytology, but population geneticists are not biochemical geneticists, and biochemical geneticists sufficiently aware of the importance of the other problems are not so common.

The department should therefore begin by providing for the teaching of population genetics and microbial genetics, if possible adding a human geneticist should the University have a Medical School. Any question of setting up a specialist department, such as of Biochemical Genetics or Human Genetics, should be viewed with disfavour, for the demand for its teaching will be inadequate to provide it with goodwill and pull in Faculties and other bodies concerned with providing funds for its development.

Thus we can lay the foundation for informed pervasive teaching by aiming at a Genetics department unspecialised in staff, separate in organisation, but integrated with other departments for its elementary teaching. It will seldom be easy as much will depend upon the personalities involved. But the stakes are high, for a great integration of biology teaching could result, and this could be far more important than the recruitment of geneticists, which will in any case look after itself, either at the Ph.D. level or, when possible, for a special honours genetics course.

I hope the scheme I have put forward here will meet with approval. Its key merit is that it allows for the fact that genetics is both part of biology and transcends the boundaries of biology. I believe that similar schemes could be applied with profit to many other University disciplines. This symposium however is concerned with genetics teaching not University teaching as a whole.

REFERENCE

ASHBY, E., CREW, F. A. E., DARLINGTON, C. D., FORD, E. B., HALDANE, J. B. S., SALISBURY, E. J., TURRILL, W. B., WADDINGTON, C. H. 1936. Genetics in the Universities, 1936. *Nature, Lond.*, **138**, 972–973.

BACTERIA AND BACTERIOPHAGES

R. C. Clowes
M.R.C. Microbial Genetics Research Unit,
Hammersmith Hospital, London, W.12

The development of microbial genetics as one of the most fundamental of the biological sciences owes much of its impetus to the fact that it forms a natural bridge between sciences as diverse as genetics and physical chemistry, biochemistry and bacteriology. It has thus encouraged workers with a wide variety of backgrounds to apply their specialised techniques and knowledge. Because of this and perhaps because of the rapidity of its growth, and also the apparent unorthodoxy of some systems, its techniques, in particular, tend to be regarded as too esoteric for the classroom. This view is perhaps, in some instances, justified, but there remain a wide variety of experiments making use of bacteria and their viruses, the bacteriophages, which are relatively easy to perform, and certainly no more difficult than those using other micro-organisms such as Neurospora and Aspergillus, which are more widely employed.

As demonstration material in teaching genetics, bacteria and phage can offer many unique features. It is of course their small size and rapid division that facilitates the accumulation and handling of large populations, while the variety of selective techniques which may be employed permits a survey of these populations for rare mutational and recombinational events with very little time and effort. This rapidity of growth is part of its appeal as teaching material, since most experiments can be completed in two or three successive days, allowing a sequential development of a topic to its conclusion. Moreover, the several operations necessary for one experiment can be delayed over longer periods, provided refrigerated storage space is available, which permits integration of experiments within most time tables. Finally, most bacterial strains can be kept viable for many years, without subculture or refrigeration using paraffin-wax sealed tubes containing a stab-culture of the strain in soft agar (Hartman *et al.*, 1960). Phage stocks can be kept for many years in buffer at 4° C. without any great loss in titre (Adams, 1959).

The intent of this paper is to outline some simple experiments with bacteria and phage which can be carried out with a minimum of apparatus, and which are illustrative either of general genetic concepts or of principles so far novel to these systems, but which are likely to have far-reaching implications. The basic apparatus required for most experiments, apart from such glassware as petri dishes, pipettes and test-tubes, is an incubator (37°) and a sterilising oven (160–200°). For some experiments a thermostatic waterbath and a small centrifuge may be necessary. Media can be purchased at all stages of preparation, ranging from individual chemical components and biological extracts to petri dishes already poured with specific media and ready for use. For most purposes, however, media can be conveniently and economically prepared from the individual constituents and sterilised by the use of a domestic pressure cooker, obviating the requirement for expensive autoclaves. Sterile plastic disposable petri dishes can sometimes be used with advantage.

The choice of suitable experiments is obviously dictated by the syllabus and the interests of the teacher. Those cited below are chosen for their elegance and for their general interest to most students of genetics.

Fine genetic structure in the T4rII phage system

The system of Benzer (1961) undoubtedly reveals the greatest degree of detail of small genetic regions, and appears to be more than sensitive enough to expose the rarest of mutational and recombinational phenomena. It depends upon the simple observation that certain mutants of phage T4, called *r*II, can readily be isolated and distinguished from the wild-type (r^+) by the larger plaques which they produce on the normal bacterial host, *E.coli* B, and by their inability to yield plaques on another strain of *E.coli* K12, (K) on which the wild-type plaques normally. If two independent *r*II mutants are crossed by mixed infection of *E.coli* B, the occurrence of one wild-type recombinant can be detected among many millions of parental particles by plating the progeny on *E.coli* K.

To carry out such a cross, about 10^9 particles of both *r*II mutants are added to a growing culture of about 10^8 *E.coli* B per ml., growing at 37° in nutrient broth. A majority of the bacterial cells then become mixedly infected with both types of *r*II mutant phage. The phages can multiply within this bacterial host cell; recombination occurs, and finally, within less than an hour, the cells burst, each releasing several hundred phage particles. A sample (0·1 ml. at a dilution of 10^{-4} to 10^{-7}) of this population is now mixed with 0·1 ml. of an overnight culture of K bacteria, and a greatly diluted sample (0·1 ml. of 10^{-8} dilution) with B bacteria, both in tubes of 2·5 ml. soft agar held molten at 45°, which are then poured over petri dishes of solid media. After a few minutes this soft agar overlay has set and the plates are then incubated overnight. The next day, a count of the plaques on K reveals the number of r^+ recombinants, and on B the number of r^+ and *r*II non-recombinant progeny. Recombinant fractions as low as 1 in 10^5 have been shown in this system, although the method can detect as few as 1 in 10^8 recombinants. Thus any degree of recombination, no matter how rare, will be detected. Lack of recombinant formation can thus be regarded as being due to the identity of the sites of the mutations involved.

Mapping with multisite mutants in the T4rII system

Well over a thousand *r*II mutants have been isolated and mapped by Benzer. In a few per cent of these (multisite mutants), the mutational damage appears to extend over a segment of the *r*II region, as judged by the inability of these mutants to give recombinants with a series of other *r*II mutants which themselves produce recombinants when crossed in any combination. It is thought that the damage results from a genetic deletion. The extent and location of these deletions in each multisite mutant have been precisely mapped. By the suitable choice of appropriate multisite mutants it is possible to map an unknown *r*II mutational site within a very small segment of the *r*II region, on the assumption that no recombination will be given with any multisite mutant in which the deletion covers the mutational site of the unknown mutant. For example, using the multisite mutants as shown in Fig. 1 it is possible to map an unknown site within one of the seven small segments A1 to B as indicated in Table I.

TABLE I

Recombination between 'multisite' and 'rII 'point' mutants

'Multisite' mutant	'Point' mutant *r* 1011	*r* DAP56	*r* G171	*r* F16	*r* 548	*r* G16	*r* 326
r 1272	o	o	o	o	o	o	o
r 1241	+	o	o	o	o	o	o
r J3	+	+	o	o	o	o	o
r PT1	+	+	+	o	o	o	o
r PB242	+	+	+	+	o	o	o
r A105	+	+	+	+	+	o	o
r 638	+	+	+	+	+	+	o
Segment	A1	A2	A3	A4	A5	A6	B

'+' and 'o' indicate respectively the presence and absence of r^+ recombinants.

The experiment is carried out by infecting a series of growing cultures of B (*c.* 10^8 cells/ml.) in small tubes at 37° by adding one drop of a suspension (*c.* 10^9 particles/ml.) of one of the multisite mutants, followed by the addition of one drop of the suspension of unknown *r*II mutant to each tube. After 45 minutes'

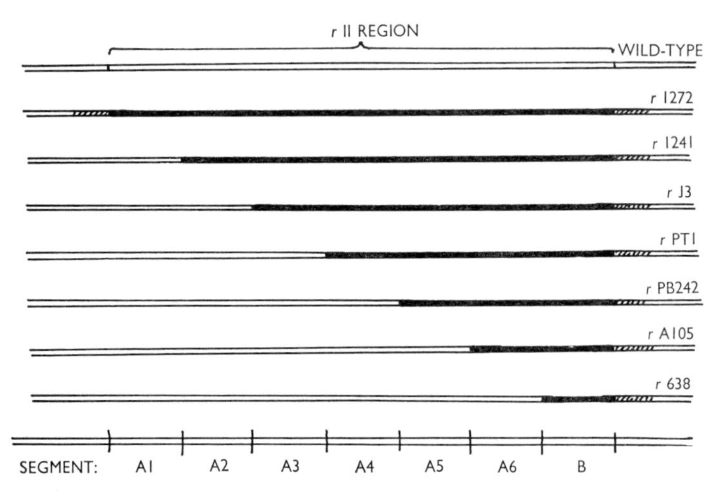

FIG. 1. *r*II region in coliphage T4 showing extent and location of 'multisite' mutants. The shaded segments indicate the material which may be deleted, and which may extend into the hatched regions.

incubation, one drop of the lysate in each tube is spotted on a petri dish of medium previously overlaid with soft agar containing K. The presence of a zone of lysis (or plaques) on this indicator strain after overnight incubation indicates recombination, with the production of r^+ phage.

Complementation in the rII *systems*

Although *r*II mutants are defined by their inability to plaque on K, lysis is sometimes found to occur if this strain is *mixedly* infected with two different *r*II mutants. The *r*II mutants can be sub-divided into two groups, *A* and *B*, on this basis, mixed infection with mutants from different groups leading to lysis, while cells infected with mutants from the same group remain unlysed. The mutational sites of each group are located together on different but adjacent regions of the phage chromosome and constitute two functionally complementary groups. One aspect of genetic mapping of *r*II mutants is to detect the complementation group to which they belong, thus allocating the site of the mutation to genetic region *A* or *B*. This can be done by a simple spot test.

About 10^7 cells of an overnight culture of K (0·1 ml.) and about 10^8 particles of the unknown *r*II mutant are added to 2·5 ml. soft agar held molten at 45° which is then poured over a nutrient agar plate. When the agar has set, spots from a suspension of an *r*II*A* and an *r*II*B* mutant (each about 10^9 particles/ml.) are placed on the soft agar and allowed to soak in. The plate is then incubated overnight. If the unknown mutant belongs to group *A*, the plate will be overgrown with a lawn of K cells, except where the spot of the *r*II*B* mutant was placed. In this region there is a clear area due to lysis of K cells which have been mixedly infected with *r*II*A* and *B* phage particles between which complementation and subsequent multiplication has occurred. Similarly, if the unknown is an *r*II*B* mutant, a bacteria-free zone will occur where the *r*II*A* drop was placed.

The spontaneous origin of mutation in bacteria

Several experiments are available to demonstrate this rather fundamental genetic premise. These include the 'fluctuation test' of Luria and Delbrück (1943), which contrasts the variation in frequency of mutants among a series of small independent populations with the lack of variation among equivalent samples from a large population. The 'replica plating' technique of Lederberg and Lederberg (1952) allows a direct demonstration of the selection of mutants without exposure of the culture to a selective agent. Unfortunately, this experiment often requires five or six successive operations at daily intervals for a successful demonstration, although each operation is short. Perhaps the simplest and most elegant method is that of Newcombe (1949).

Aliquots (*c.* 2×10^7 cells) of an overnight culture of *Escherichia coli* are spread over a series of nutrient agar plates. The plates are then incubated for $3\frac{1}{2}$ hours to allow several generations of growth, so that each bacterial cell produces a small clone. Half of the plates are now re-spread with buffer to separate and re-distribute the cells within the clones, while the other half remain unspread. All the plates are then sprayed with T1 phage (*c.* 10^{11} particles/ml.) so that the entire surface of the plate is covered, without allowing any droplets to form. In this way, all but resistant cells are killed without disturbing the bacterial distribution on the plate. After overnight incubation it is found that the spread plates show about a twenty-fold increase in the numbers of T1 resistant colonies compared to the numbers on those plates which were unspread, indicating the clonal nature of resistance to phage T1. Induction of resistance by the phage itself predicts that the number of T1 resistant colonies would be proportional only to the size of the population treated, which should therefore give the same number on both series of plates.

Chemical mutagenesis

The use of chemical mutagens has been greatly extended in recent years. Many mutagenic chemicals are, however, toxic, and may require specific conditions for optimal activity. The use of ethyl methane sulphonate (EMS), first used with T2 phage (Loveless, 1958) and later on with bacteria (Loveless and Howarth, 1959), offers a quick-acting, non-toxic and highly mutagenic agent ideally suited for the demonstration of induced mutagenesis.

A vigorously growing culture of an auxotrophic bacterial strain derived by dilution of an overnight culture 1/20 in broth, and reincubating for 1½ hours at 37°, is centrifuged and concentrated by resuspension in a small volume of 0·4M EMS and incubated for 15 minutes at 37°. The culture is then diluted tenfold in buffer, centrifuged and washed to remove EMS and spread over a minimal-glucose medium lacking the required growth factor. On this medium only prototrophic revertants can grow to produce clones. It is found that the number of revertants may be increased one hundred- to a hundred-thousandfold, depending on the mutation studied. Increases from 1 in 10^8 (spontaneous) to about 1 in 10^6 after EMS are typical with *Salmonella typhimurium LT2, cysA21* (Loveless and Howarth, 1959); with *E.coli W945*, reversions from *thr*⁻ to *thr*⁺ increase from 1 in 10^{10} to 1 in 10^5 (Loveless—personal communication).

Oriented genetic transfer in Escherichia coli *K-12*

The *E.coli* K-12 mating system can be used to demonstrate genetically the linear arrangement of markers on a genetic structure. In K-12, matings occur by a one-way genetic transfer from one parent (donor) to another parent (recipient). In this transfer, the genome of an Hfr-type donor parent is transferred to the recipient in a linear oriented manner, at a constant rate, starting in all cells from the same chromosomal location. The transfer of any particular gene into the recipient does not therefore occur until a fixed time after the contact of the parents. If at varying times after mixing, the parental cells are separated, the chromosome breaks and only those genes already transferred can appear in the recombinants. By this means the times of entry of different genes can be measured. By the choice of Hfr donors differing in the point of entry and direction of transfer, it can be shown that the distances between any genes measured by the time of entry is constant (Jacob and Wollman, 1958).

The demonstration system uses a high-frequency donor strain such as Hfr 'Reeves 1' which is prototrophic (*thr*$^+$, *leu*$^+$, *met*$^+$) and streptomycin sensitive (*str-s*) mated with a recipient (F^-) strain which requires threonine, leucine and methionine for growth (*thr, leu, met*) and is streptomycin resistant (*str-r*). Equal volumes (5 ml.) of rapidly growing cultures of both parents (*c* 2×10^8/ml.) are mixed and kept at 37°. At five-minute intervals (up to 40 minutes) small (0·2 ml.) samples are taken, diluted tenfold into buffer and shaken vigorously for about half a minute. This shaking is best done on a mechanical shaker such as a 'Microid' shaker, but can be performed satisfactorily either by the use of an eccentrically mounted rubber

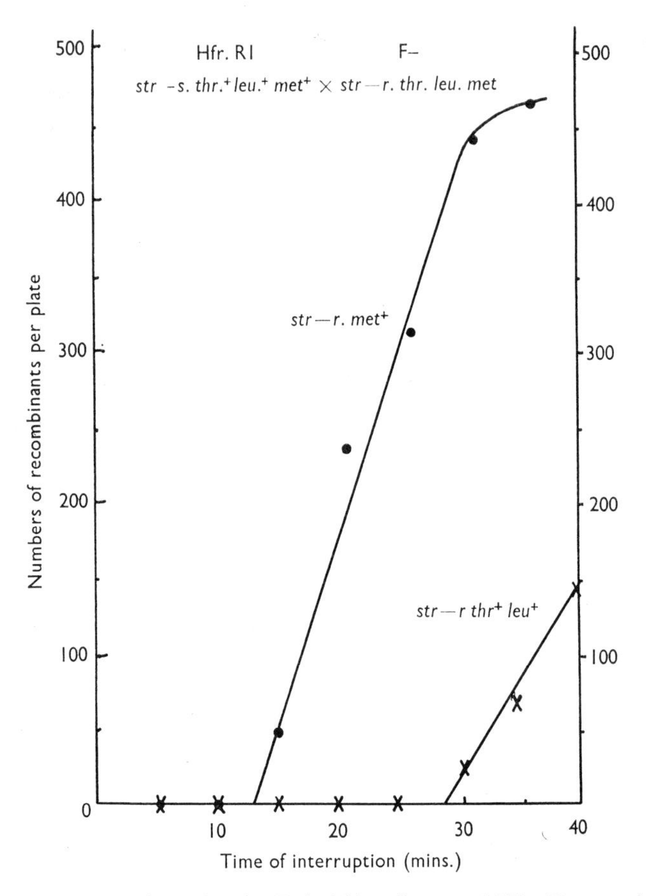

FIG. 2. Interrupted mating in *Escherichia coli* K-12 of Hfr (Reeves 1) *str-s thr + leu + met+* with W1F (*str-r thr leu met*) at 37°.

bung on a geared axle, or even by vigorous hand shaking. Aliquots are then diluted 1:100 in buffer, and 0·1 ml. samples spread over minimal agar glucose plates plus streptomycin, containing either threonine and leucine or methionine. After overnight incubation, the number of *str-r met+* and *str-r thr+ leu+* recombinant colonies on these plates can be plotted against the time of interruption. By extrapolation, the time of entry of the genes *met*$^+$ and *thr*$^+$ *leu*$^+$ can be measured. Typical results are shown in Fig. 2.

APPENDIX—GROWTH MEDIA

1. Nutrient Broth

Oxoid No. 2 nutrient broth powder	25 g.
Water	to 1000 ml.

Dissolve the broth powder. Sterilise at 15 lb. for 15 mins. (pH7·5 approx.).

2. Nutrient Agar

Oxoid No. 2	25 g.	Suspend agar and broth powder in water. Steam at 100° until dissolved (15 litres takes about 1½ hrs.). Dispense into screw-cap bottles. Sterilize at 15 lb. for 20 mins. (pH7·4 approx.).
Davis New Zealand Agar	12·5 g.	
Water	to 1000 ml.	

3. Minimal Agar

(*a*) *Minimal medium salts* (x4)

NH_4Cl	20 g.	Dissolve each salt in cold water in the order indicated, waiting for complete solution until adding next. (A light precipitate is formed.) Filter into 25 ml. or 100 ml. bottles. Sterilise at 15 lb. for 10 mins. (No further precipitate should be formed.) pH7·2.
NH_4NO_3	4 g.	
Na_2SO_4 (Anhydrous)	8 g.	
K_2HPO_4	12 g.	
KH_2PO_4	4 g.	
$MgSO_4.7H_2O$	0·4 g.	
Water	to 1000 ml.	

(*b*) *Water agar* (x4/3)

Davis NZ Agar	20 g.	Suspend agar in water and steam at 100° until dissolved. ADJUST pH TO 7·2. Dispense in 75 ml. or 300 ml. bottles. Sterilise at 15 lb. for 20 mins.
Water	to 1000 ml.	

(*c*) *Minimal agar*

Water agar (x4/3)	300 ml.	Melt agar at 100° or under pressure (15 lb. for 15 mins.). Add warmed salts and glucose. Mix and dispense into plates.
Salts (x4)	100 ml.	
*D-Glucose (20%)	4 ml.	

4. Soft Agar Overlay

As Water Agar (3b) except using 6 g. agar per litre. No nutrients are necessary.

* Make up separately and sterilise at 5 lb. for 10 mins.

5. Buffer

(*a*) Minimal medium salts (after appropriate dilution) is an adequate buffer for most purposes. Dilute with sterile (15 lb. for 15 mins.) distilled water.

(*b*) If a less complex buffer is required,

KH_2PO_4	3 g.	Dissolve in order. Sterilise at 15 lb. for 10 mins.
Na_2HPO_4 (Anhydrous)	7 g.	
NaCl	4 g.	
$MgSO_4.7H_2O$	0·2 g.	
Water	to 1000 ml.	

6. Supplements

(*a*) Carbohydrates are usually added to 0·2%.
(*b*) Amino-acids are usually added to 20 μg/ml. (L or DL).
(*c*) Vitamins are usually added to 1 μg/ml.

REFERENCES

ADAMS, M. H. 1959. *Bacteriophages.* Interscience Publ. Ltd., London.

ANAGNOSTOPOULOS, S. AND SPIZIZEN, J. 1961. Requirements for transformation in *Bacillus subtilis. J. Bact.*, **81,** 741.

BENZER, S. 1961. On the topography of the genetic fine structure. *Proc. Nat. Acad. Sci., Wash.*, **47,** 403.

HARTMAN, P. E., LOPER, J. C. AND ŠERMAN, D. 1960. Fine structure mapping by complete transduction between Histidine—requiring Salmonella mutants. *J. gen. Microbiol.*, **22,** 354.

JACOB, F. AND WOLLMAN, E. L. 1958. Genetic and physical determinations of chromosomal segments in *Escherichia coli. Symp. Soc. exp. Biol.*, **12,** 75.

LEDERBERG, J. AND LEDERBERG, E. 1952. Replica plating and indirect selection of bacterial mutants. *J. Bact.*, **63,** 399.

LOVELESS, A. 1958. Increased rate of plaque type and host-range mutation following treatment of bacteriophage *in vitro* with Ethyl Methane Sulphonate. *Nature, Lond.*, **181,** 1212.

LOVELESS, A. AND HOWARTH, S. 1959. Mutation of bacteria at high levels of survival by Ethyl Methane Sulphonate. *Nature, Lond.*, **184,** 1780.

LURIA, S. E. AND DELBRÜCK, M. 1943. Mutations of bacteria from virus sensitivity to virus resistance. *Genetics*, **28,** 491.

NEWCOMBE, H. B. 1949. Origin of bacterial variants. *Nature, Lond.*, **164,** 150.

SPIZIZEN, J. 1958. Transformation of biochemically deficient strains of *Bacillus subtilis* by deoxyribonucleate. *Proc. Nat. Acad. Sci., Wash.*, **44,** 1072.

TRANSDUCTION IN BACTERIA

E. A. Bevan
Botany School, Oxford University

An experiment on transduction can be used for teaching many important principles of bacterial genetics. If the whole operation is carried out in the teaching laboratory it requires three stages and these can be on successive days or weeks to fit a teaching time-table.

Principles

We speak of transduction when genetic material from one bacterium, the donor, is transmitted to another, the recipient, by the agency of a bacteriophage. Zinder and Lederberg in 1952 first demonstrated transduction using strains of *Salmonella typhimurium* and the temperate phage PLT22. Temperate phages differ from virulent ones in that a fraction of a bacterial population survive their infection. In these the phage becomes established as an immature form, the prophage. Such bacteria are termed lysogenic (Bertani, 1958). They are immune to re-infection by the same type of phage and give clones of prophage-carrying cells which rarely lyse spontaneously but which may be induced to lyse and produce mature phage if treated with ultra-violet light (Lwoff, 1952).

The use of temperate phage both permits the recovery of phage progeny from infected donor cells, and it allows a fraction of infected recipient cells to become lysogenic and survive. Should any recipient cell be infected by a phage particle carrying genetic material from the donor, then transduction may follow. Such transduced bacteria may be lysogenic or non-lysogenic. The lysogenic result follows infection simultaneously with a transducing phage and a normal phage; the non-lysogenic follows infection from a transducing phage alone. With the temperate phage PLT22 non-lysogenics may be found also among the first-formed progeny of a simultaneously infected cell in which the prophage has not yet become established and is therefore unable to divide in step with the bacterial genome (Luria *et al.*, 1958). In *Salmonella* (but perhaps not in other bacteria) a transducing phage does not lysogenise: presumably, this is because a part of the normal phage genome is replaced by the genetic material which it carried from the donor bacterium (see Clowes, 1960).

Two general properties of transduction should now be noticed. First the frequency of transduction of gene markers from one bacterium to another is low—generally of the order of 1 in 10^4 infected recipient bacteria. Secondly, the amount of donor genetic material carried by a single transducing phage is small, and, consequently, it rarely happens that two independent gene markers are transduced simultaneously. If they happen to be closely linked, however, they both may be included in a single particle and joint transduction may follow its entry into a bacterium.

It sometimes happens that a genetic fragment may be transferred into a recipient bacterium but not become integrated into its genotype. Under these conditions *abortive* transduction takes place (Stocker, 1956; Ozeki, 1956). When a recipient is abortively transduced from auxotrophy to prototrophy the event is manifested by the appearance of a tiny colony following prolonged incubation on the selective agar medium used to detect transduction. The small size of the colony is due to the fact that the fragment, which contains the gene conferring prototrophy, is able to persist but is unable to duplicate. This means that at any one time in colony development only one cell is dividing—that possessing the fragment.

Strains of bacteria and phage

Many strains of phage and bacteria are available to illustrate transduction. In the experiment described below a stock suspension of temperature phage PLT22 derived from the LT2 wild-type strain of *Salmonella typhimurium* (the donor) is mixed with a population of recipient bacteria containing the two linked mutant genes leu-39 and ara-9 (leucine requiring, unable to ferment arabinose). Both wild-type and mutant strains of bacteria were obtained originally from the Cold Spring Harbor Collection, New York. By the procedure described below the numbers of these bacteria which become jointly transduced ($leuc^+$, ara^+), and singly transduced for the leucine marker ($leuc^+$, ara^-), may be determined.

Experimental Procedure

The procedure is set out to suit weekly practical classes. For the benefit of the teacher a list of sterile materials required per student or student group for each class precedes the protocol of each part of the experiment. Unless otherwise stated the composition of media is the same as that given by Clowes in the preceding article.

1st Week: Transduction

Materials

(a) Overnight broth culture of recipient bacteria leu-39 ara-9 containing approximately 2×10^9 cells per ml. prepared by inoculating a small loopful of bacteria into 10 ml. of nutrient broth in a 'bubbler' tube and incubating with aeration at 37°C.

(b) Both wild type and homologous phage preparations (i.e. built up in wild-type and ara^- leu^- bacteria respectively) at titres of $2–3 \times 10^{10}$ per ml.

(c) One 10 ml. pipette; fourteen 1 ml. pipettes; two screw-top phials, and two centrifuge tubes.

(d) Nineteen dishes of nutrient agar; eleven dishes of enriched minimal agar medium (minimal medium according to Clowes formula to which has been added 1 per cent. nutrient broth). Because *Salmonella typhimurium* LT2 is a motile bacterium these dishes should be poured a few days before the experiment to allow the agar to dry out a little.

(e) Seven screw-top phials each containing 9·9 ml. of T2 buffer (Na_2HPO_4 Anhyd. 7 grm.; KH_2PO_4, 3 grm.; NaCl, 5 grm.; $MgSO_4$ 0·1 M, 10 ml.; $CaCl_2$ 0·1 M, 10 ml.; 1 per cent. gelatine, 1 ml.; water, 980 ml.).

Protocol

1. To obtain a washed suspension of recipient bacteria, centrifuge the broth culture for 10 minutes at 3500 r.p.m. Decant the supernatant, and resuspend the cells in an equal volume of T2 buffer.

2. Mix 2 ml. of washed recipient bacteria and 2 ml. of wild-type phage suspension in a screw-top phial and stand for 8 minutes to allow for absorption of the phage.

3. Plate 0·1 ml. aliquots of the mixture on to 5 enriched minimal medium dishes. Distribute the suspension evenly over the surface of the agar with a glass spreader. These dishes will allow bacteria transduced to leucine independence to grow and produce colonies.
4. As a control, repeat operations 2 and 3 using homologous phage.
5. Spread 0·1 ml. aliquots of washed recipient bacteria on 3 enriched minimal medium dishes. These dishes will give the frequence of spontaneous reversions at the leucine locus.
6. To obtain the titre of the washed bacteria, dilute by 10^{-6} and spread 0·1 ml. aliquots on 5 nutrient agar dishes. Make these dilutions in the usual way—0·1 ml. into 9·9 ml. of buffer—three times using a fresh pipette for each transfer.
7. To obtain the titre of the surviving bacteria in each mixture repeat operation 6.
8. To check that the phage preparations are free from bacteria, spread 0·1 ml. aliquots of each on 2 nutrient agar dishes.
9. Label the dishes and incubate for 36 hours at 37°C., after which they may be transferred to a refrigerator until the following week.

2nd and 3rd Weeks: Scoring results and testing for joint transduction

Materials

(a) Five dishes of eosin methylene blue nutrient agar with arabinose. (This is an indicator medium on which colonies of bacteria able to ferment arabinose have a distinctive greenish sheen whereas non-fermenters are pink. Its composition is: 2·55 per cent. nutrient agar, 0·04 per cent. eosin Y, 0·006 per cent. methylene blue, and 1 per cent. arabinose.)

(b) A replica plating apparatus as described by Lederberg (1952).

Protocol

1. From last week's dishes (a) calculate the percentage of surviving bacteria which have been transduced to leucine independence, and (b) record the frequency of spontaneous reversions at the leucine locus.
2. To determine the percentage of joint transductions (i.e. leu^+ ara^+) replica plate each of the dishes containing leu^+ transduced colonies on the EMB—arabinose dishes.
3. Label the dishes and incubate overnight at 37°C. Next morning, transfer the dishes to a refrigerator.
4. In the 3rd week score the leu^+ colonies for their ability to ferment arabinose, and calculate the percentage of joint transductions among the total transductions at the leucine locus. These should be in the region of 65 per cent. (Smith-Keary, 1960).

Precautions

Salmonella typhimurium is a pathogenic bacterium. During the course of these experiments, therefore, every precaution should be taken against infection. All pipettes should be discarded in lysol solution. All petri dishes and tubes containing bacteria should be discarded on trays which can then be transferred to the autoclave. Dishes containing bacteria should not be opened unless it is really necessary: colonies should be counted without opening the dishes by marking the bottom of a dish with a wax pencil. Should the laboratory bench become contaminated, alcohol may be used to disinfect the area. Students should wash their hands as often as possible during the course of the experiment.

REFERENCES

BERTANI, G. 1958. Lysogeny. *Advance in Virus research,* **5,** 151–193.

CLOWES, R. C. 1960. Fine genetic structure as revealed by transduction. *Symp. Soc. Gen. Microbiol.,* **10,** 92–114.

LEDERBERG, J. AND LEDERBERG, E. N. 1952. Replica plating and indirect selection of bacterial mutants. *J. Bact.,* **63,** 399.

LURIA, S. E., FRASER, D. K., ADAMS, J. N. AND BURROWS, J. W. 1958. Lysogenisation transduction, and genetic recombination in bacteria. *Cold Spring Harbor Symp. Quant. Biol.,* **23,** 71–82.

LWOFF, A. 1952. The nature of phage reproduction. *Symp. Soc. Gen. Microbiol.,* **2,** 149–163.

OZEKI, H. 1956. Abortive transduction in purine—requiring mutants of *S. typhimurium.* In *Genetic studies with Bacteria, Carnegie Inst. Wash.,* Pub. 612, 97–106.

SMITH-KEARY, P. F. 1960. A supressor of leucineless in *Salmonella typhimurium. Heredity,* **14,** 61–71.

STOCKER, B. A. D. 1956. Abortive transduction of motility in Salmonella. *J. Gen. Microbiol.,* **15,** 575–598.

Review articles

ADAMS, MARK H. 1959. Bacteriophages. *Interscience,* New York.

HARTMAN, P. E., AND GOODGAL, S. H. 1959. Bacterial Genetics with particular reference to genetic transfer. *Ann. Rev. Microbiol.,* **13,** 465–504.

FUNGAL GENETICS

J. A. PATEMAN
Department of Genetics
Milton Road, Cambridge

There are two good reasons for using fungi as practical material for teaching genetics. They can provide elegant demonstrations of many aspects of genetics, e.g. linkage, and mutation, with equal facility to that afforded by organisms such as Drosphila. In addition they afford examples of phenomena such as: segregation in single tetrads, heterocaryosis and somatic recombination which are difficult or impossible to demonstrate in other organisms. A brief outline will be given of some experiments which have been used in practical classes. This will include an indication of the possible numbers and level of students for which the experiment is suitable and any specialised equipment required.

Growth tests of biochemical mutants

Fresh cultures of two or three different biochemical mutants of *Neurospora crassa* are required. There is a large number of mutants with single specific requirements available. One culture of each type for every 3–4 students is sufficient. Each student is given test tubes variously containing sterile water, minimal liquid medium and minimal medium supplemented with one of the specific requirements of the mutants. A clump of spores is shaken in the sterile water and the tubes of media inoculated with drops of spore suspension on a sterile loop. The tubes are incubated at 25° or room temperature 18–20° if necessary, and after 2–3 days growth of mycelium is observed, but only in tubes with an appropriate supplement. This demonstrates that each mutant has single specific chemical requirements for normal growth. It is suitable for any number of first year university students and requires only test tubes, inoculating loops, and petri dishes if combined with a demonstration of heterocaryosis.

Heterocaryons

In conjunction with the growth tests, the formation of heterocaryons may be demonstrated with biochemical mutants. Small clumps of spores are placed on the surface of solid minimal media contained in petri dishes. Single inocula of mutant strains produce no growth on incubation of plates at 25°. If clumps of spores from genetically different strains are superimposed, growth will often occur due to the formation of complementary heterocaryons. Heterocaryons between spore colour mutants of *Aspergillus* are dealt with in another section.

Segregation and linkage

Strains of *Neurospora crassa* of different mating type grown together in a mating tube produce perithecia which discharge haploid ascospores on to the wall of the tube. A clump of several thousand ascospores may be scraped off the tube wall and washed off in a drop of sterile water on the surface of agar medium in a petri dish. The spore suspension is spread evenly with a bent glass rod and the plate kept in an oven at 60°C. for 30–60 minutes. After subsequent incubation at 25°, normal ascospores produce hyphae 1–2 mm. long after about 20 hours. Ascospores carrying a mutant allele determining a biochemical requirement usually germinate on minimal medium to give a very short hypha, but do not grow further and are quite distinct from the normal. A cross between a biochemical mutant and wildtype will give 1:1 segregation in the ascospores. A cross between two unlinked mutants will give 3 slow growers to 1 normal. If two linked biochemical mutants are crossed an estimate of the map distance between the loci is given by twice the percentage of normal ascospores. In this way segregation and linkage can be demonstrated in populations of thousands of individuals. This experiment is suitable for first or second year students, but requires a binocular dissecting microscope with at least × 40 magnification, for each student.

Backmutation

A culture of a biochemical mutant strain of *Neurospora crassa* grown in a test tube on appropriate medium will produce up to 10^8 conidia. The conidia are washed off the slope and passed through a loose cotton wool filter to provide about 10 ml. of a suspension containing $5–10 \times 10^6$ conidia per ml. The spore suspension is irradiated in an open petri dish with ultra-violet light for a period sufficient to kill 99 per cent. of the spores. The irradiated suspension is centrifuged and the spore pellet resuspended in 2 ml. of water. A 0·1 ml. aliquot of the suspension is spread on each of 20 petri dishes containing minimal solid medium with 1·5 per cent. sorbose added. The sorbose induces a colonial growth in *N. crassa*. The plates are incubated at 25°; after 2–4 days small colonies will be observed. These backmutant colonies can be picked off, and cultured. A growth test will show that they can grow without the specific supplement required by the original strain. This experiment is suitable for third year students. It requires an ultra-violet lamp, or other mutagenic agent, and a centrifuge, and the number of students would normally be limited to a maximum of about twelve.

Segregation in Tetrads

An ascospore colour mutant of the fungus *Sordaria macrospora* produces pale coloured instead of the normal black ascospores. *S. macrospora* is a homothallic fungus, but if the normal and the mutant strains are cultured together some hybrid perithecia are formed. A suitable method of producing hybrid perithecia is to blend together cultures of the mutant and normal strains which have been grown on agar slopes. The mixed cultures are added to agar medium at 50°C. which is then poured into petri dishes to set. After 5–10 days at 25° large numbers of perithecia will be formed. About 10 per cent. of these perithecia will be hybrid and contain asci segregating 1 : 1 for pale coloured and black ascospores. If the perithecia are broken by squeezing with fine tweezers in a drop of 50 per cent. glycerine on a slide then the asci, each of which is a single tetrad, are readily observable. Permanent preparations can be made by mounting the asci in glycerine jelly or in Euparal after dehydration. From the permanent preparations of hybrid asci the students can count the frequencies of first and second division segregation and thus determine the map distance of the ascospore colour locus from the centromere. Alternatively the students can make their own preparations if more time is available. There is also an ascospore colour mutant of *N. crassa* known which can be used in a similar fashion. This material is suitable for any number of first year or second year students.

The arginosuccinase system in Neurospora crassa

Strain B317 is a mutant which requires arginine for normal growth. It is known that this mutant lacks the enzyme arginosuccinase which catalyses the following reaction:

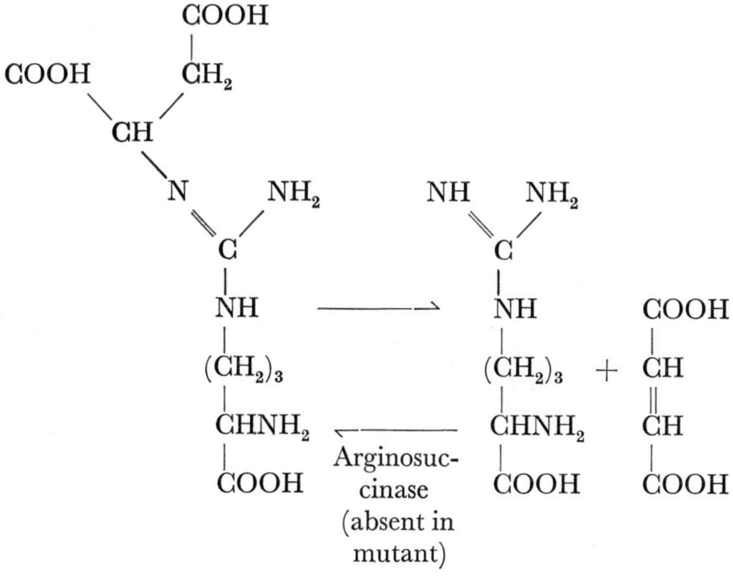

COOH
CH_2
COOH
CH
N NH$_2$
C
NH
$(CH_2)_3$
$CHNH_2$
COOH

⟶ / ⟵ Arginosuccinase (absent in mutant)

NH NH$_2$
C
NH
$(CH_2)_3$
$CHNH_2$
COOH

\+

COOH
CH
CH
COOH

Arginosuccinic acid (accumulated in mutant) *Arginine* *Fumaric acid*

(i) *Accumulation of precursor.* An arginineless strain is grown in Fries minimal + M/400 L-arginine HCl medium and in Fries minimal + M/400 L-arginine HCl + M/400 citrulline for 72 hours at 25°C. A wildtype strain is grown under similar conditions on Fries minimal and on similar supplemented media to the *arg* strain. The mycelium is harvested, washed on a Buchner funnel, blotted dry and weighed. The mycelium is placed in equal weight of buffer and extracted in a boiling water bath for 5 minutes. The pads are squeezed out against side of tube with a clean glass rod, leaving amber coloured extract. Three superimposed spots of extract are run on a Whatman No. 1 paper chromatogram, the solvent is phenol saturated with water and with ammonia in atmosphere. The mutant extracts give spots of arginosuccinic acid with R_F ca 0·3, just ahead of glutamic acid R_F ca 0·25. There is no arginosuccinic acid spot in any of the wildtype extracts. The spots are developed by spraying the chromatogram with 0·25 per cent. ninhydrin in *n*-butanol after the solvent has been dried off.

(ii) *Demonstration of arginosuccinase.* Wildtype and *arg* are grown for 48 hours at 25°C. in culture bottles each containing 150 ml. Fries minimal or Fries minimal + M/400 L-arginine HCl. The mycelium is harvested, washed, blotted dry and weighed. Then ground for 5 minutes in ice-cold mortar in × 10 blotted weight of M/50 pH 7·4 phosphate buffer, centrifuged and the supernatant dialysed 2–3 hours with stirring against two changes of the same buffer. Then 0·6 ml. 0·05 M Na fumerate, 0·6 ml. 0·05 M L-arginine HCl, 0·2 ml. 0·02 MpH 7·4 phosphate buffer and 0·6 ml. enzyme extract incubated together for 1 hour at 37·5°C. Also control system with Na fumerate replaced by phosphate buffer. Reaction stopped by heating 5 minutes in boiling water bath. Then three superimposed spots run on chromatogram similar to (1). Arginosuccinic acid R_F ca 0·3, arginine ca 0·85–0·9. This experiment is suitable for up to twelve third year students at a time.

The Parasexual cycle

The various stages of the parasexual cycle in *Aspergillus nidulans* can be demonstrated using strains such as the following:

(1) su_1ad_{20}, ribo-1, an-1, prol-1, paba-1, y, ad_{20}.

(2) w_2, ad_{20}, bi-1.

The symbols represent the following characteristics: ribo-1 = requirement for riboflavine; an-1 = requirement for anurine; prol-1 = requirement for proline; paba-1 = requirement for p-aminobenzoic acid; y = yellow conidia; ad_{20} = requirement for adenine; w_2 = white conidia; bi-1 = requirement for biotin; su_1ad_{20} = suppressor of ad_{20}. ad_{20} is a leaky allele at the ad_8 locus and its suppressor su_1ad_{20} is recessive so that both the heterocaryon and the diploid between these two strains have a partial requirement for adenine. w_2 (white conidia) is epistatic to y, but both colour mutants are recessive. The biochemical requirements are recessive.

The diploid between the strains (1) and (2) will be green and have a partial requirement for adenine but not for any of the other supplements. Selection from this diploid, of yellow segregants (homozygous for y) and adenine independent segregants (homozygous for su_1ad_{20}), makes it possible to ascertain the order of the mutants on chromosome I and also the position of the centromere.

The following is a brief outline of the various stages; further details can be obtained from the account by Pontecorvo *et al.* (*Advances in Genetics*, 1953). Some or all of the stages can be used for class purposes depending on the time available; no elaborate equipment is necessary.

Inoculate a thick suspension of conidia from both strains on to the surface of liquid complete medium contained in a small vial, mixing the two types of conidia thoroughly. After a few days' growth a mycelial mat will have formed on the surface of the liquid medium. Collect the mycelial mat, and wash two or three times in sterile distilled water. Tease out pieces of the mycelium on the surface of a plate of minimal medium supplemented with adenine. Vigorously growing heterocaryotic hyphae will emerge from the mycelium after a few days' growth. Transfer blocks of conidia from the leading edge of such a heterocaryotic sector to plates of minimal supplemented with adenine. The heterocaryon will bear a mixture of white and yellow conidial heads. Pour about 5 ml. of dilute detergent solution (Tween 80) on to a plate with conidiating heterocaryotic colonies. Scrape off the conidia with a sterile loop and pipette off the resulting conidial suspension into a vial. Separate the conidia by sucking them up and down vigorously for 50–100 times in a Pasteur pipette with a capillary spout. Filter the resulting suspension through a cotton wool plug into a centrifuge tube, centrifuge down the conidia and resuspend the pellet in 1 ml. of dilute detergent solution. Do a haemocytometer count of this suspension to ascertain the total number of conidia being plated. Then spread 0·1 ml. of the conidial suspension on each of ten plates of minimal medium supplemented with adenine. Only green prototrophic diploid colonies will grow on the minimal medium, apart from the occasional heterocaryotic colony which, however, will not have green conidia and conidiate less densely. Streak diploid colonies on to plates of minimal medium plus adenine. Restreak diploid colonies on plates of minimal medium plus adenine. Scrape off small amount of conidia from a diploid colony and place in a drop of detergent solution on a slide and cover with cover-slip. Measure the conidial size using micrometer eyepiece at magnification $\times 400$ and a sample of 20–50 conidia. Do the same with known haploid and diploid cultures for comparison. The diploid conidia should have a volume about twice that of the haploid conidia and show an increase in diameter by a factor of about $\sqrt[3]{2} = 1{\cdot}26$.

For selection of yellow and white segregants make a dilute suspension of conidia in detergent solution and separate as above. Do a haemocytometer count and arrange to plate 10–20 conidia on each of five plates of complete medium. Also inoculate the dilute conidial suspension at nine marked positions on each of two plates containing complete medium. Yellow and white segregants from the resultant diploid colonies can be picked off as appropriately coloured conidial sectors under a binocular microscope at a magnification of $\times 30$. These should be streaked on to marked positions on plates of complete medium. To purify segregants restreak to marked positions on plates containing complete medium. All segregants should be tested for biochemical requirements by making point inoculations on a series of plates lacking in turn just one of the requirements, and on complete medium as a control. Classify segregants for biochemical requirements by scoring for growth or no growth on each plate.

THE USE OF YEAST FOR TEACHING PRACTICAL GENETICS

E. A. Bevan
Botany School, Oxford
R. A. Woods
Department of Genetics, University of Sheffield

INTRODUCTION

Because the meiotic products of fungi are sampled and classified directly they are more suitable for demonstrating some fundamental genetic principles than the gametes of higher organisms which, with few known exceptions, must be fertilised before their genotype can be determined. Here we are going to describe the experiments on yeast breeding carried out by both the elementary and advanced students at the Botany School, Oxford. They are designed to illustrate the principles involved in meiotic and mitotic recombination, polyploid inheritance, interallelic complementation, and cytoplasmic inheritance.

Our choice of yeast (*Saccharomyces cereviseae*) was made for the following reasons: 1. It has both a haploid and, unlike most fungi, a diploid stage in its life-cycle, and sexual reproduction is easily controlled; 2. Stable mutant strains are available and simple to isolate; 3. It is a unicellular organism which, like many bacteria, produces distinct compact colonies on agar medium and does not produce asexual airborne spores; and 4. It is suitable for demonstrating all the principles mentioned above.

Most of the basic techniques and media used in yeast genetics have been reviewed by Winge and Roberts (1958) and Emerson (1955). For more detailed information we shall refer the reader to original literature. Those techniques which have been evolved in the course of our work but which have not yet been published will be described more fully.

STRAINS

Two morphological mutant strains, ad_1 and $ad_{2\cdot0}$, are used in our experiments. Each requires adenine for growth and produces an intracellular red pigment. Their colonies are therefore easily distinguished from normal white yeast (see illustration in Lindegren, 1949). We also use a number of mutant auxotrophic strains. These are se_1, his_8, me_2, which require serine, histidine, and methionine for growth respectively. For the experiments demonstrating cytoplasmic inheritance we breed the vegetative (*veg*) and segregational (p_1) 'petite' strains described by Ephrussi (1953). A list of mutant strains and a genetic map of yeast is given by Hawthorne and Mortimer (1960).

EXPERIMENTS

A. The production of heterozygotes and asci

The mass mating technique of Lindegren (1949) is the easiest for students to carry out. Here, actively growing haploid cells of opposite mating type are inoculated into a tube of nutrient medium and incubated at 28°C. Fusion bridges and the large diploid cells can be seen in a few hours when examined under a microscope. We usually cross different auxotrophic strains and following their mating streak the contents of the tube on minimal medium where only the prototrophic diploid cells grow. Alternatively, advanced students cross individual haploid cells by placing them side by side on a nutrient agar plate with a micromanipulator and isolating the diploid cell which forms within 3 to 4 hours. This technique is a modification of that described for isolating spores of hymenomycetes by Kemp and Bevan (1959).

Diploid cells transferred to acetate agar medium (Fowell, 1952) and incubated at 25°C. will form asci in 2 to 3 days.

For most of our experiments students are supplied either with the diploid heterozygous strain or the asci already formed from it.

B. Random spore analysis

Recently we have developed a method for random spore analysis of yeast which, unlike tetrad analysis (Winge and

Laustsen 1937), and the analysis of intact asci (Bevan, 1953), enables even the most inexperienced student to analyse the progeny from a given heterozygote. Using this method we obtain ascospores free of unspored vegetative cells by adding medicinal paraffin to a mixed aqueous suspension of them and shaking vigorously. The ascospores which are hydrophobic remain in the paraffin layer whereas the hydrophilic vegetative cells do not. The details of our method follow.

A suspension is prepared containing approximately 10^7 asci per ml. (haemocytometer count) in 10 ml. of a 10 per cent. solution of snail gut enzyme. This dissolves the ascus wall but the spores are unaffected (Johnston and Mortimer, 1959).* After 4 hours' incubation at 28°C., a few free spores will be present but most tend to stick together in spite of the absence of the ascus wall. These are separated by centrifuging the solution, discarding the supernatant and resuspending the cells in 1·5 ml. of sterile water. This is then ground in a Potter homogeniser, either by hand or with the aid of an electric motor, until the solution becomes milky owing to some of the cells breaking down and releasing their contents.

The homogenised solution in which most of the cells are now separate is made up to 10 ml. with sterile water in a screw-cap universal container, 10 ml. of medicinal paraffin is then added and the container shaken well for 5 minutes, following which the two layers are allowed to settle out.

The paraffin layer is then washed four times by removing the aqueous layer with a Pasteur-pipette, adding 10 ml. of fresh sterile water and shaking well.

The ascospores may be counted in a haemocytometer using a phase contrast microscope, or serial dilutions followed by spreading 0·05 ml. amounts on 2 per cent. nutrient agar may be made to determine the titre. Dilutions are best made by shaking 1 ml. of suspension with 9 ml. of fresh paraffin over water. We aim at obtaining a final titre of 3×10^3 spores per ml. which gives approximately 150 colonies per plate when 0·05 ml. is spread.

The suspensions may be kept in a refrigerator at 2°C. for many months without loss of viability. We find that less than 2 per cent. of the colonies arising from our suspensions are diploid.

In Table I we have listed the genotypes of the strains and agar media employed in our experiments, and where possible, the theoretically expected results.

C. Tetrad analysis

Students are always impressed by the demonstration of a 1 : 1 segregation of alleles among spores from a hybrid ascus.

* We prepare our enzyme from the common garden snail, *Helix aspersa*. The stomach contents are dissected out, 1 ml. of water added per stomach, and the solution seitz-filtered. This is diluted with sterile water to a 10 per cent. solution.

TABLE I

Breeding experiments with yeast (for details see text)

Principle illustrated	Diploid genotype*	Growth medium	Ratio of colonies expected Red	: White
Independent assortment	$ad_{2\cdot0}$ +/+ me_2	complete	1	1
		minimal + adenine		
Linkage	$ad_{2\cdot0}$ +/+ se_1**	complete	1	1
		minimal + adenine	4 (20 units)	1
Epistasis	$ad_{2\cdot0}$ +/+ ad_6***	complete	1	3
Gene interaction	$ad_{2\cdot0}$ +/+ ad_1	,,	3	1
Mitotic recombination	$ad_{2\cdot0}$ his_8/+ +	,,	—	
Interallelic complementation	$ad_{2\cdot0}/ad_{2\cdot1}$	,,	—	
	$ad_{2\cdot1}/ad_{2\cdot2}$ etc.			
			Normal	: Little
Cytoplasmic inheritance	+ × p_1	,,	2	2
	+ × + (veg)	,,	4	0
	p_1 × + (veg)	,,	2	2

* Reciprocal genotypes are also analysed.

** For a three-point test-cross his_8 situated 16 units from se_1 is used.

*** ad_6 controls a step in adenine synthesis prior to that controlled by ad_2. All strains possessing it are white.

Yeast asci derived from a red-white heterozygote are suitable for this purpose but must be dissected with a micromanipulator. This is facilitated by the method of Cox and Bevan (1961) or by that of Hawthorne and Mortimer (1960), but is too time consuming for the elementary student to undertake. We therefore find that asci of *Sordaria fimicola* heterozygous for a gene determining spore colour are more suitable for this demonstration (Olive, 1956). Here the ascospores are ordered and the distance of the colour gene from the centromere can be calculated directly (Catcheside, 1951). Further, these asci may be mounted on slides and used year after year.

D. Mitotic recombination

Mitotic recombination in yeast may be increased by ultraviolet irradiation (James and Lee-Whiting, 1955). In our experiments we irradiate the $ad_{2\cdot0}$ his_8/+ + heterozygote in 9 ml. of saline at a titre of 10^4 cells per ml., giving a dose which allows 10 per cent. of the cells to survive. Following mitotic recombination a proportion of colonies arising from these cells are either pure red or sectored red-white. A sample of red cells from each colony is transferred to a fresh plate of complete agar medium and the resulting growth replica plated (Lederberg and Lederberg 1952) on minimal medium plus adenine to determine what percentage of the isolates do not require histidine. This percentage is a measure of the frequency of recombination between the adenine and histidine genes.

E. Interallelic complementation

The red strains of yeast are ideal material for demonstrating interallelic complementation. Its occurrence is detected by the white colour of the heterozygous diploids derived by crossing pairs of independently isolated red ad_2 haploid mutant strains.

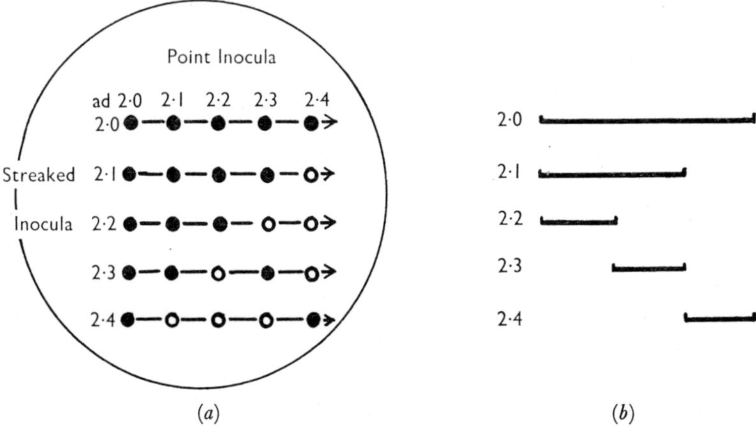

FIG. 1. (*a*) The method of testing the complementation relationships of red mutants on complete agar medium. The inocula are heavy suspensions of cells in saline. Streaked inocula are of opposite mating type to point inocula. Clear circles indicate the white growth of complementing diploids, and full circles the red growth of non-complementing diploids.

(*b*) The linear complementation map derived from these relationships.

Thus there is no need for testing the heterozygotes on minimal medium as in the cases described in *Neurospora* (Catcheside and Overton, 1958). Further, the complementation patterns of many mutants can be tested on the same dish of nutrient agar. So far we have isolated 21 *ad* strains ($ad_{2\cdot 0}$ . . . $ad_{2\cdot 20}$). When tested in pairwise combinations they fall into 10 groups and a linear complementation map of the locus shows 5 sub-units (Bevan and Woods, 1962). We supply stocks of five of these strains to our students for testing according to the method illustrated in Fig. 1.

F. Polyploid inheritance

Polyploid strains may be produced according to the method of Roman *et al.* (1954). The asci of appropriate strains when dissected yield phenotypic ratios of 4 : 0, 3 : 1 and 2 : 2. Our advanced students use the tetraploid $ad_{2\cdot0}$ $ad_{2\cdot0}$ + +/+ + me_2 me_2 and study the red : white segregations from its asci.

G. Cytoplasmic inheritance

Ephrussi and his colleagues (for a review see Ephrussi, 1953) have shown that among the colonies arising from a plated suspension of either haploid or diploid yeast cells there are 1–2 per cent. whose diameter is only one-third to one-half that of the rest. The frequency of these 'vegetative' littles or 'petites', as they are called, may be increased to almost 100 per cent. by treatment of the cells with euflavine. These 'vegetative' littles were shown to be due to a cytoplasmic rather than a nuclear mutation. Later, it was demonstrated that the same phenotype may be bestowed by a nuclear mutation and the colonies arising from the cells possessing this mutation were called 'segregational' littles. As well as other differences from normal, each type of mutant lacks the respiratory enzymes cytochromes a and b, cytochrome oxidase, and succinic dehydrogenase.

We utilise both types of mutants to demonstrate cytoplasmic inheritance to our classes. The crosses performed are given in the Table. In addition to the crosses thick suspensions in water of both normal and mutant cells are examined under a Hartridge reversion spectroscope where differences in spectra in the cytochrome region can be seen clearly.

REFERENCES

BEVAN, E. A. 1953. Genetic analysis of intact incomplete asci of yeast. *Nature*, **171**, 576–577.

BEVAN, E. A., AND WOODS, R. A. Complementation between adenine requiring mutants in yeast. *Heredity*, **17**, 141

CATCHESIDE, D. G. 1951. *The Genetics of Micro-organisms*. Sir Isaac Pitman and Sons.

CATCHESIDE, D. G., AND OVERTON, ANNE. 1958. Complementation between alleles in heterocaryons. *Cold Spring Harbor Symp. Quant. Biol.*, **23**, 137–140.

COX, B. S., AND BEVAN, E. A. 1961. A new technique for isolating spores of yeast (*Saccharomyces cereviseae*). *Trans. Brit. mycol. Soc.* **44,** 239–242.

EMERSON, S. 1955. Biochemical genetics. *Handbuch der physiologisch-und pathologisch-chemischen Analyse,* **10** Aufl., Bd. **11,** 443–537.

EPHRUSSI, B. 1953. *Nucleo-cytoplasmic Relations in Micro-organisms.* Clarendon Press, Oxford.

HAWTHORNE, D. C., AND MORTIMER, R. K. 1960. Chromosome mapping in *Saccharomyces*: centromere linked genes. *Genetics,* **45,** 1085–1110.

JAMES, P., AND LEE-WHITING, B. 1955. Radiation-induced genetic segregations in vegetative cells of diploid yeast. *Genetics,* **40,** 826–831.

JOHNSTON, J. R., AND MORTIMER, R. K. 1959. Use of snail digestive juice in isolation of yeast spore tetrads. *J. Bact.,* **78,** 272.

KEMP, R. F. O., AND BEVAN, E. A. 1959. New Techniques for isolating spores of hymenomycetes. *Trans. Brit. mycol. Soc.,* **42,** 308–311.

LEDERBERG, J., AND LEDERBERG, E. M. 1952. Replica plating and indirect selection of bacterial mutants. *J. Bact.,* **63,** 399–406.

LINDEGREN, C. C. 1949. *The Yeast Cell, its Genetics and Cytology.* Educ. Pub. Incorporated, St. Louis.

OLIVE, L. S. 1956. Genetics of *Sordaria fimicola.* I. Ascospore color mutants. *Amer. J. Bot.,* **43,** 97–107.

ROMAN, H., PHILIPS, M. M., AND SANDS, S. M. 1955. Studies on polyploid *Saccharomyces.* I. Tetraploid segregation. *Genetics,* **40,** 546–561.

WINGE, Ø., AND LAUSTSEN, O. 1937. On two types of spore germination and on genetic segregations in *Saccharomyces*, demonstrated through single spore cultures. *Ctes rend. Lab. Carlsberg, Ser. Physiol.,* **22,** 99–116.

WINGE, Ø., AND ROBERTS, C. 1958. Yeast genetics. In *the Chemistry and Biology of Yeasts,* 123–156. Edited by A. H. Cook. Academic Press, New York.

BIOMETRICAL GENETICS

J. L. Jinks

A.R.C. Unit of Biometrical Genetics, Department of Genetics
University of Birmingham

Content of Course

Biometrical genetics is the study of continuous variation and as such it is both an analytical and an experimental science. The analytical approach requires both a knowledge of Mendelian genetics and that part of mathematical genetics concerned with the construction of models that can account, as adequately as possible, for the processes involved in hereditary transmission of continuously varying characters. The experimental approach requires in addition an understanding of the principles underlying the designs of the special types of experiment used in the study of continuous variation and of the statistical manipulation of the data from such experiments.

Biometrical genetics is, therefore, largely a combination and extension of Mendelian genetics and biometry and hence it is appropriately taught only after a good grounding has been given in these two principal components. At Birmingham all biology students are given such a grounding, at an elementary level, in both Mendelian genetics and biometry. The former is taught in a first year course consisting of 24 hours of lectures and 40 hours of practicals, the latter in a second year course consisting of 20 hours of lectures and 15 hours of tutorials. These courses jointly provide a basis for the subsequent teaching of biometrical genetics although they are not designed primarily for this purpose. They are general comprehensive courses for students who do not wish to specialise in genetics and as such they include a treatment of biometrical genetics at an appropriately elementary level. At the same time they are preparatory courses for those students who wish to specialise in genetics in their third year. Hence these courses must prepare the latter for advanced treatments of chromosome mechanics, gene theory and population genetics as well as biometrical genetics.

The third year course in biometrical genetics consists of 10 hours of lectures, that is 13 per cent. of the total lecturing time, and appropriate practicals spread throughout the year; taking approximately one third of the total of 70 days devoted to practical work. It starts from the assumption that the third year students are familiar with Mendelian genetics and those parts of biometry concerned with the estimation of means, variances, covariances, correlations and regressions, tests of significance (c, t, variance ratio, χ^2) and the principles of experimental design. Additional statistical techniques as well as extensions of those listed are required during the course, for example, multiple regression and maximum likelihood estimation of parameters, and these are taught as the need arises during the practical classes which accompany the course.

The 10 hours of lectures which are devoted to biometrical genetics are divided between the various topics as follows.

1 hour. The history of biometrical genetics

1 hour. Polygenic inheritance

1½ hours. Scales and scaling tests

2½ hours. The additive, dominance and environmental components of generation means, variances and covariances following:

(i) A cross between two inbred lines continued by selfing, sibing and backcrossing.

(ii) Crosses between members of populations that are inbred, outbred with and without Hardy equilibria, and with and without equal gene frequencies.

2½ hours. Modifications arising in the above situations and specification of effects due to:

(i) Non-random distribution of genes and linkage

(ii) Epistasis

(iii) Genotype-environment interaction

1½ hours. Illustrations of uses and applications of biometrical genetics. For example, for investigating the genetical basis of heterosis and for predicting the speed and limits of selection.

Both the approach and the content of this course are based on Mather's book *Biometrical Genetics* (1949) although its content has been augmented by new illustrative material and by treatments of diallel crosses, epistasis and genotype-environment interaction which have appeared during the past decade.

In the practical course attempts are made to illustrate many of the issues raised in the lectures. The practicals are of four types.

1. The application of biometrical analysis to plant breeding data. Suitable data for this purpose are always available from the past and present research programmes of the department. The use of these data has the advantage that the students have themselves seen and assisted with these, or similar, experiments in the experimental field in a previous summer vacation course.

2. An experimental investigation of the inheritance of a continuously varying character. For example, the parent, F_1, F_2 and backcross generations of a cross between two inbred lines of *Drosophila melanogaster* may be raised and scored for sternopleural chaeta number. The design, execution of this experiment and the subsequent analysis of the generation means and variances will illustrate most aspects of biometrical genetics.

3. Selection for high and low expression of a continuously varying character. For example, selection for high and low sternopleural chaeta number following a cross between two inbred lines of *D. melanogaster*. If the inbred lines used are the same as those in experiment 2 the students may compare the estimate of heritability from the latter with the realised heritability which is obtained from a comparison of the observed and expected advance under selection.

4. A comparison of different experimental designs. For example, a randomised block and latin square design may be compared for plants grown in a glasshouse subject to environmental gradients. Our students have compared *Nicotiana rustica* varieties under these conditions using the length of the first and second true leaves as the character.

These formal courses are supplemented throughout the year by essays and seminars on various biometrical problems and by discussions between the students and the staff and research students who are engaged in research in biometrical genetics.

A Special Problem

Difficulties arise, of course, in teaching biologists a subject which requires some mathematical ability. These spring from the relative inability of the average biology student to cope with mathematics. This in turn leads to a dislike or even resentment of all subjects that require some understanding of mathematics. The problem appears to be initiated two or three years before the student enters a University. At this stage many students take up biology or even botany and zoology as separate subjects to avoid taking the more mathematical sciences and mathematics itself. Others in deciding to specialise in biology find that they are thereby debarred from taking mathematics, the two subjects often being strict alternatives in the school syllabus. This situation may be further aggravated by the Universities themselves whose requirements for entry to an honours school in biology often discourage the student from taking mathematics as a principal or subsidiary subject either before or after entering the University. The net outcome is a negative correlation between the desire to enter an honours school in biology and the ability to do mathematics, part of which is inherent and part the result of training. In consequence no more than half the biology students who attend the course in biometry reach a level of competence which is adequate for a proper understanding of biometrical genetics. Thus the problem in teaching biometrical genetics is largely the problem of teaching biometry. Indeed, in our experience, those students who can understand the latter have few difficulties with the former.

While we have discussed this problem in relation to the teaching of biometrical genetics it is not confined to this subject. In a less acute form it is met with wherever a predominantly experimental science is taught to biologists.

Conclusions

Because of its novelty and relatively high mathematical content biometrical genetics is one of the more difficult branches of genetics to teach successfully. Furthermore, it is an experimental science and adequate time must be set aside for the rather lengthy practical experiments which are necessary to illustrate the lectures. At the same time the importance of teaching biometrical genetics cannot be overemphasized. Not only is it a main tool for the genetic improvement of plants and animals but it offers an experimental approach to population genetics and is essential to the understanding of the factors involved in natural selection and evolution. No advanced course in genetics is complete or balanced without biometrical genetics any more than it would be without chromosome mechanics or physiological genetics.

REFERENCE

MATHER, K. 1949. *Biometrical Genetics.* Methuen, London.

Acknowledgement: I am indebted to Professor K. Mather, C.B.E., F.R.S., for helpful discussion.

A PRACTICAL EXERCISE IN QUANTITATIVE GENETICS

J. M. Thoday
Department of Genetics
Milton Road, Cambridge

The practical exercise to be described here has three purposes. First, it complements practicals in biometrical genetics by showing how relatively precise information can be obtained concerning the location of some of the genes affecting a classical quantitative character, sternopleural chaeta number in *Drosophila melanogaster*. Second it provides first hand data for simple statistical analysis. Third it gives the student an insight into the ways in which marker genes may be used to manipulate chromosomes in breeding programmes. The principles upon which the exercise is based are described by Thoday (1961).

The exercise involves three breeding programmes which are designed to demonstrate the approximate location in the chromosomes of the genes that are mainly responsible for the high chaeta number of flies heterozygous for genomes from the selection line dp-1 of Thoday and Boam (1961). Four stocks are required: dp-1, *y bw st*, *st p*, and *se cp e*. Binoculars giving at least 30× magnification are needed. Whether the students make the crosses themselves or are merely given the various progenies must depend on the timing of the practicals.

First the students count chaetae in the four stocks, and the reciprocal F_1s dp-1 × *y bw st* and *y bw st* × dp-1. They thus get practice in counting, information concerning the chaeta numbers of the stocks and demonstrate that the dp-1 × chromosome has little effect on chaeta number.

Second the students count chaetae in equal numbers of each sex and marker genotype in the progeny of the testcross *y bw st* ♀ × *y* dp-1/*bw* dp-1/*st*. Typical results are given in Table I from which it can be seen that the relevant genes lie in chromosome III.

Third the students count the chaetae in equal numbers of each sex in each of the marker genotypes in the progeny of the testcross dp-1/*st p* ♀ × *st p*. Typical results are given in Table II showing that the relevant genes lie in the left arm of chromosome III.

Table I

Results of the *y bw st* testcross

Genotypes	++	*bw* +	+ *st*	*bw st*
Chaetae per fly	23·825	24·100	20·700	21·250

Table II

Results of the *st p* testcross

Genotypes	++	+ *p*	*st* +	*st p*
Chaetae per fly	23·250	23·250	20·825	20·775

Table III

Results of the *se cp e* testcross

Genotypes	+	*se cp e*	+ *cp e*	*se* ++	++ *e*	*se cp* +	+ *cp* +	*se* + *e*
Chaetae per fly	20·59	16·43	18·43	18·64	20·83	16·61	17·86	18·70

Fourth the students count the chaetae in equal numbers of each sex and genotype in the progeny of the testcross dp-1/*se cp e* ♀ × *se cp e*. Typical results are given in Table III. The *se cp* recombination classes have intermediate and similar chaeta numbers. Three hypotheses fit this observation:

1. There is one chaeta number locus approximately equidistant between *se* and *cp*.
2. There are two or more chaeta number loci between *se* and *cp*.
3. There are two chaeta loci one quasi-allelic to or to the left of *se* and the other quasi-allelic to *cp*.

Fifth, to distinguish between these hypotheses, the students progeny test the *se* ++/*se cp e* and + *cp e*/*se cp e* males by further testcross to *se cp e*. They choose flies for testcross from the top, the middle and the bottom of each of the two distribution curves given by the *se* ++ and + *cp e* flies. They find that the three classes of flies give significantly different progenies and it follows that there are at least two high chaeta number loci between *se* and *cp* in dp-1.

The number of flies that can be scored depends upon the number of students and the time available. Five flies of each sex and genotype in each exercise is ample if there are 5 students, except that extra *se* ++/*se cp e* and + *cp e*/*se cp e* males are required in order to give an adequate distribution for selection of the males for progeny testing. If there are fewer students more time is needed for each student must count more flies.

The whole exercise can be much improved by using an inbred wild-type stock in parallel with dp-1 for this enables the chaeta number differences distinguishing the marker stocks to be subtracted out. The *se cp e* stock for instance has an unusually low chaeta number largely due to the presence in it of a low chaeta number factor in the *cp* region. Inbred Oregon was one of the ancestors of dp-1 and its use adds interest to the exercise. The introduction of the inbred wild-type standard, however, doubles the work involved. Reduction of the work can be made by eliminating some of the stages. In fact the *se cp e* testcross with progeny testing is all that is essential. However the results of this cross make much more impact, and hence the principles get over much better, if the simpler crosses precede it.

dp-1 (and the other stocks if necessary) can be made available to any University teacher who wishes to maintain it in order to lay on practicals along these lines.

REFERENCES

THODAY, J. M. 1961. The location of polygenes. *Nature, Lond.*, **191**, 368–370.

THODAY, J. M. AND BOAM, T. B. 1961. Regular responses to selection. 1. Description of responses. *Gen. Res.*, **2**, 161–176.

THE USE OF MICE IN TEACHING GENETICS

D. S. FALCONER
Institute of Animal Genetics, Edinburgh

Many people find mice attractive animals, and an interest in genetics may often be more readily evoked by mice than by other more convenient though more remote organisms. The segregation of coat colours displayed in a litter of mice makes an impressive demonstration of Mendel's laws, and mice provide excellent material for the simultaneous classification of several segregating genes. The idea that an array of phenotypes can be reduced to a number of simple 'either-or' alternatives is one of the most difficult for the beginner to grasp. The repeated presentation of the problem of multiple classification in successive litters of mice at intervals of some weeks gives time for a real understanding of this essential idea to develop.

Segregation and Factor Interactions

Two genes very suitable for the demonstration of Mendelian ratios, of simultaneous classification and of gene interaction are non-agouti (*a*) and brown (*b*). Both are autosomal recessives, unlinked, and affect coat colour though in different ways. They are widely distributed among mouse stocks and can be obtained from pet shops, where the double homozygote (*aa bb*) is known as 'chocolate'. The wild-type, which is perhaps less readily obtained from pet shops, is known as 'agouti'. Detailed descriptions of the two genes are given by Grüneberg (1952). Their effects on pigmentation are as follows:

Non-agouti affects the distribution of pigment in the hairs, and brown affects the colour of the pigment. The wild-type hair has granules of black pigment throughout its length except in a band—the agouti band—near the tip, which contains yellow pigment. This gives the coat a grey and speckled appearance. The non-agouti mutant obliterates the yellow band by extending the black pigment throughout the whole of the hair. In the absence of other colour mutants this gives a uniformly black coat.

The brown mutant changes the black pigment to brown without affecting the yellow pigment in the agouti band. In the absence of other mutants this gives a brown speckled coat, known as 'cinnamon'. Brown and non-agouti together give a uniform brown coat known as 'chocolate'.

Thus the simultaneous segregation of the two mutants produces four distinctive coat colours—'grey' (i.e. wild-type), 'black' (*aa*), 'cinnamon' (*bb*), and 'chocolate' (*aabb*). These descriptive names for the colours, however, are a frequent source of confusion in the interpretation of the genetics, and it is of the greatest importance that the student should be taught to classify the two genes independently; $+^a$ or *aa* by the presence or absence of agouti bands, and $+^b$ or *bb* by the black or brown colour of the pigment.

Mice obtained from pet shops may well carry some other colour genes and these, though complicating the classifications, can add to the interest of the interactions. (i) Albino (*c*) is the most likely to be found. It is easily recognised by the complete lack of pigment in the coat and eyes. It illustrates the meaning of epistasis because no other colour genes can be classified in albino homozygotes. (ii) Chinchilla (c^{ch}) is an allele of albino. Its chief effect is on the yellow pigment of the agouti band which is rendered colourless. It slightly dilutes the black pigment in the rest of the hair but has no appreciable effect on the brown pigment of *bb* homozygotes. (iii) Pink-eye (*p*), a mutation at a different locus, has an effect opposite to that of chinchilla: it does not affect the yellow pigment but changes the black pigment to a pale grey-brown colour. It also removes the pigment from the eye and, like albino, it can be classified at birth by the absence of pigment in the eye. The interaction of pink-eye and chinchilla is interesting. Since one affects the black pigment and the other the yellow, the two in combination remove almost all the pigment from the hairs and produce a phenotype almost indistinguishable from albino.

Pink-eye and chinchilla are linked, with a recombination

(or crossing over) frequency of 16 per cent. in segregating females and 12 per cent. in males. Unfortunately, however, the two genes are not readiiy available in coupling. The linkage, therefore, could be demonstrated only in a repulsion F_2, unless coupling homozygotes were prepared well in advance. There are many other linked genes including several sex-linked ones, but suitable combinations are not readily available.

Developmental Genetics

Another purpose for which mice are very well suited is to follow the embryological development of anatomical defects. Two genes that are good for morphogenetic studies are Danforth's short-tail (*Sd*) and Brachyury (*T*). Both are autosomal dominants that are lethal in homozygotes. The phenotypic effects in heterozygotes are superficially similar—reduction in the length of the tail—but the morphogenetic causes are different.

Action of Sd

The shortening of the tail of *Sd* heterozygotes and homozygotes follows a secondary degeneration of the notochord and is caused by haemorrhage with a constriction and peripheral degeneration of the tail. Homozygotes are more severely affected than heterozygotes. The constriction occurs nearer the base of the tail and there are defects of the urinogenital system leading to an imperforate anus and often a lack of one or both kidneys. The absence of the kidneys is interesting because it results from a failure of the induction of the metanephros by the ureter. A further point of interest about *Sd*/+ heterozygotes is that the odontoid process of the axis vertebra is missing. The axis-atlas joint is modified structurally and there is no apparent loss of function.

Action of T

The *T*-defect follows a primary defect of the development of the notochord. In *T*/+ heterozygotes the defect is confined to the distal part of the tail, which develops as a thin filament that later disappears. But in *T*/*T* homozygotes the defect is much more severe and the whole posterior half of the body fails to develop.

The two genes illustrate several points of genetical interest: (i) the development of a mutant starts normally and then diverges along an abnormal pathway so that the mutant becomes progressively more unlike the normal: (ii) the abnormalities of heterozygotes and homozygotes are similar in nature, but the abnormality of heterozygotes appears later and is consequently less severe; (iii) dominance differs in degree—*Sd* shows intermediate dominance while *T* is nearly a recessive lethal; (iv) lethal genes kill by a specific defect of development and the missing phenotypic class can be found if looked for.

Class Work

The genetics and embryology of the two mutants are fully described by Grüneberg (1952, 1953, 1958). For the following summary of class work I am much indebted to Mrs. R. M. Clayton. The features mentioned can all be seen with a low power binocular microscope, and a prior knowledge of embryonic morphology is not necessary.

Matings should be made between heterozygotes and the litters should first be examined at birth. The new-born litters of *Sd*/+ × *Sd*/+ matings contain all three phenotypic classes. Most of the *Sd*/*Sd* homozygotes are found alive at birth, but they die within the first day. Their defects can be seen to be more severe than those of the heterozygotes, and they can be identified by the imperforate anus.

The shortening of the tail in *T*/+ heterozygotes is sometimes very slight and may amount to no more than a blunting of the tip. Careful classification, however, will give a segregation ratio of 2:1 in the litters of *T*/+ × *T*/+ matings, indicating the absence of the homozygote class at birth; later work will reveal the fate of the missing homozygotes.

When the specific abnormalities have been recognised at birth, embryos can be examined. The embryos are best examined fresh. They may be fixed, for comparison with embryos examined later, but the morphological features are less clear in fixed material. The chief interest centres round the 9–11 day stages, but it is best to start with embryos of 14 or 15 days, and work backwards to earlier stages. The older embryos are easier to handle than the younger, and their defects are easier to recognise. The earliest stage that can easily be examined without section-cutting is 9 days. The abnormalities of both *Sd/Sd* and *T/T* homozygotes appear first at this stage, or a little earlier, but both the heterozygotes are still indistinguishable from the normal embryos. The visible abnormalities from 9 days onwards are as follows.

At 9 days the *Sd/Sd* homozygotes have haemorrhages in their tails, but the *Sd/+* heterozygotes are still externally normal. At 10 days heterozygotes have haemorrhages and homozygotes are more severely abnormal. At 11 days and onwards there are varying degrees of abnormality and the homozygotes can no longer be readily distinguished from heterozygotes.

The *T/T* homozygotes at 9 days are much more severely defective than the *Sd/Sd* homozygotes. They are still living but they completely lack the posterior part of the body. At about 11 days the *T/T* homozygotes die, and the *T/+* heterozygotes begin to be distinguishable from the normal embryos. After 11 days the homozygotes are represented by the remains of the degenerating placentae which persist throughout the pregnancy. The *T*-litters from 11 or 12 days onwards therefore give a 1:2:1 ratio, representing dead homozygotes, heterozygotes, and normal homozygotes, respectively. In the later litters, of 14 or 15 days, the degenerating placentae are quite small—about 2 mm. in diameter—and can easily be missed unless the uterus is examined after removal of the living embryos.

Technical Aspects

Mice can be kept in almost any sort of wooden or metal cage. They need some sawdust or granulated peat, and paper shavings, wood wool, or hay for bedding. (Cotton wool is unsuitable because it forms ligatures round the legs of the young mice.) Commercial mouse foods are available, but dog biscuits or a mixture of grains will suffice. Water must be provided.

The minimum generation interval is about nine weeks. Gestation takes about twenty days. The young mice can be weaned at three weeks of age, and they are sexually mature at about six weeks.

Copulation, which usually takes place at night, can be diagnosed by the presence of a hard 'plug' in the vagina. If the plug is not visible externally it can be felt with a blunt seeker. Embryos can be timed from the copulation plug. Females will usually mate again within 24 hours of the birth of a litter, but thereafter not until the litter is weaned. Post-partum mating gives the highest rate of production of young.

The gestation period is prolonged to about 4 weeks when concurrent with the suckling of the young of a previous litter. Females in continuous production therefore produce litters at intervals of about 4 weeks. The number born per litter varies greatly according to the strain and the circumstances: about six might be assumed for the purposes of planning. Surplus mice are best killed with ether or coal-gas.

REFERENCES

GRÜNEBERG, H. 1952. *The Genetics of the Mouse.* Nijhoff, The Hague.

GRÜNEBERG, H. 1953. Genetical studies on the skeleton of the mouse. VI. Danforth's short-tail. *J. Genet.*, **51,** 317–326.

GRÜNEBERG, H. 1958. Genetical studies on the skeleton of the mouse. XXII and XXIII. *J. Embryol. exp. Morph.*, **6,** 124–148, 424–443.

TEACHING CYTOLOGY

H. REES
Department of Agricultural Botany
University College of Wales, Aberystwyth

Different university departments may naturally be expected to teach cytology in very different ways. A biophysics department, where emphasis is placed mainly on cell structure, may inevitably concentrate its teaching upon the physical organisation of nuclear and cytoplasmic components and upon their mechanics. Cytology as an integral part of a genetics course, and within the framework of a general biological training, is presented with very different aims in view. This matter of integrating the teaching of cytology, and for that matter genetics, with biology generally presents one of the most important problems to the teacher. It is one worth discussing before dealing with more specific issues.

Aims

In designing a course of lectures in cytology and genetics for first and second year students one recognises, as many others have pointed out, the unique opportunity offered of cutting through the traditional and often formidable barriers between the various biological teaching compartments. Apart from cytology and genetics, biochemical, taxonomic, ecological matters, even, remarkably, plants and animals together in the same context can be profitably considered. To cut through these barriers is to kill two birds, perhaps three, with one stone. First, a proper sense of unity between the various biological sciences is indicated. Second, the versatility and the power of the genetic method to inquiry in these other fields is demonstrated. Third, students may be convinced that cytology and genetics are in themselves worthy of further specialised study.

To achieve these three things would be to achieve much. Above all it requires the presentation of a course in heredity in the broadest context, to range from molecules and biochemistry to populations and evolution—a course in which the cytology and the genetics each have their separate and, more important, their combined parts to play.

It might, of course, be argued that it is not for teachers of cytology and genetics to range so widely and generally at the expense, often, of detail in strictly cytological or genetical topics. There are, however, two good reasons why they should. Firstly, the notions of genetics imparted to students at schools are, too often, restricted and old fashioned. The subject is taught in a void with no indication of its relevence to population dynamics or to evolutionary problems generally. By and large school genetics seems chiefly devoted to detailed, sometimes even inaccurate, descriptions of mitosis and meiosis, and to the memorising of a variety of ratios illustrating major gene segregations. Not surprisingly a large proportion of students are put off. Secondly, at the university level, the organisation of biological teaching is frequently such that only the geneticists and, occasionally, biochemists are unrestricted to rigid teaching compartments, especially to dealing with plants and animals separately. In future no doubt there will come into being more biology departments teaching biology.[1] Until then one is justified, perhaps even obliged, in the first two years to take advantage of the genetics and cytology to cover, albeit briefly, broader aspects of the subject. At the same time a deeper understanding of cytology and genetics should and, I think, does result.

If these aims are agreed they provide convenient terms of reference for constructing the introductory courses. Since the emphasis is on genetics and the control of heritable variation in populations the cytology becomes then mainly concerned with the material basis of heredity, in brief with:

1. The storage of biological information.
2. Its distribution.

On this theme of information and distribution the cytologi-

[1] A group of scientists discussed and reported on this matter in 1936 (*Nature, Lond.*, 138, 972). A committee of Fellows of the Royal Society has been set up recently for the same purpose.

cal processes and their consequences can be conveniently and effectively dealt with at the various levels of organisation:

1. The molecular, in terms of nucleic acids and fine structure.
2. The cellular, in terms of chromosome structure and behaviour.
3. The individual, in terms of recombination and chromosome change.
4. The population, in terms of chromosome polymorphisms, selection and adaptation.

The theme, it seems to me, provides a suitable framework for the cytology in itself and affords, at the various levels of treatment, the necessary links with, and between, other relevant fields of inquiry.

Practice

During the first two years the emphasis in teaching is properly placed on laying the general foundation, and on stating the general principles of the subject. In practical work, therefore, it is not perhaps essential or, for reasons of time, always possible to deal with matters of detail. The two important aspects are the training in elementary techniques of making cytological preparations and the provision of demonstration material illustrating the main kinds of variation in chromosome structure, number and behaviour.

Students, even large classes, can easily handle the Feulgen technique for root tips, preferably using well tried and simple material such as onion or broad bean. For meiosis one can use aceto-carmine or orcein squashes. Grasshopper testes, rye and *Tradescantia* anthers are admirable. It is, of course, fortunate that these techniques, though simple, are perfectly adequate in one form or another for most cytological investigations of chromosomes. This being so there is no advantage in devoting long periods of time to demonstrating more obscure and little used methods.

Prepared slides for demonstrating chromosome variation present a little more of a problem. It is preferable to show the consequences of interchanges, inversions and other changes in material with relatively few and reasonably large chromosomes.

It is an advantage also to show as much of the variation as possible within the one group of organisms. The more familiar the students become with the standard complement the better are they able to comprehend and appreciate the consequences of its variation. Rye is a most suitable material for this purpose. Haploids, triploids and tetraploids: interchanges and inversions can be demonstrated involving the same basic chromosome complement. For showing polyploidy *Tradescantia* is also a useful and readily available group. One other most useful source of material is *Crocus balansae*. For the whole sequence of mitosis and meiosis its chromosomes are as good as any.

Turning now to rather more advanced work for third year students, the principle of studying intensively the chromosome variation within one group is again most useful. In rye, in addition to the numerical and structural variants referred to earlier, inbred lines are available which illustrate a wide range of chromosome variation under genotypic control (Rees, 1955). This variation ranges from subtle differences in chiasma frequency and localisation to gross abnormalities involving chromosome breakage or asynapsis. Straightforward comparisons between lines are useful both from a strictly cytological standpoint as well as affording training in statistical and genetical analyses of populations in terms of cytological characters.

One other exercise with rye is worth mentioning as an illustration of work on the cytology of population. Lines are available which contain two interchanges (see Rees, 1961). These can be distinguished from one another in heterozygotes because one conveniently involves the chromosome organising the nucleolus and the other does not. The class is presented with fixcd heads from twenty to thirty plants derived from self pollinating a double interchange heterozygote. They are then asked, (1) to classify the plants in terms of interchange types, (2) to determine whether the distribution of interchanges is at random or provides evidence of selection for certain interchange combinations, (3) to estimate the variation in disjunction frequencies of interchange configurations within the population. The exercise as well as being useful is most economical in terms of effort in preparation.

To enlarge further upon the details of material for class work is hardly useful. Obviously each department will have its own convenient material available. It is perhaps just worth emphasising again that students probably do benefit from detailed work on one type. Clearly work should not be confined to it but at the same time it is best to avoid as far as possible a demonstration that shows, say, haploidy in rye, triploidy in *Tradescantia* and supernumerary chromosomes in locusts—at least for the beginners. One final point concerning teaching material. It would be useful, as many teachers have indicated, to know who has what to spare for use by other departments. During research work quite useful teaching material turns up that could often be made available without much trouble. All that is needed is a list and some goodwill.

To turn next from practical work to theory for advanced students, and in particular to the question of selection of subject matter for training. It is inevitable in Honours courses not completely devoted to genetics that the time available for lectures and discussions imposes restrictions on the scope and intensity of the genetics curriculum. Each department naturally constructs its genetic course to suit its own special needs in relation to the Honours syllabus generally. Each will have a bias towards certain aspects of the subject and the variety which results in the way of student products in the country at large is no doubt to everyone's advantage. In this respect the choice of subject matter is fairly straightforward and readily determined. The danger with shorter courses is not so much in defining the scope or extent of the course but, rather, in neglecting depth or intensity of training. One should in short try to build one's course not only with an eye to the relevance of its information but also on the basis of its value as an intellectual exercise. As an example, a detailed analysis of the theory of crossing over ranging from the cytological interpretation of chiasmata to a complete analysis of ordered tetrads is a most useful intellectual exercise in itself and one which, in addition, brings forth a whole battery of techniques to bear on the one problem in impressive testimony of the genetic method. Even at the expense of four or five lectures on some perfectly useful topic the exercise is well justified.

Finally, there is one general point to consider. It is that most students in their first year rarely display wild enthusiasm towards either genetics or cytology. It may be that their first introduction to the subject at school is the cause. Whatever it is we should go out of our way to capture their interest as much as possible. It is for their own benefit. The subject, we all agree, is interesting and all that is required is to make clear at the outset what it is, how it is applied. For this purpose an introductory lecture on the history of the science can be most useful and, withal, entertaining. To create interest in the subject in fact is a proper and respectable aim. A few years ago, at the end of one series of lectures, I asked a young lady if she had understood the work. She said no but that she found it exceedingly interesting. This, at any rate, was something.

REFERENCES

REES, H. 1955. Genotypic control of chromosome behaviour in rye. I. Inbred Lines. *Heredity*, **9**, 93–116.

REES, H. 1961. The consequences of interchange. *Evolution*, **15**, 145–152.

MATERIAL FOR DEMONSTRATING CHROMOSOME BEHAVIOUR

K. R. Lewis
Department of Botany, University of Oxford
B. John
Department of Genetics, University of Birmingham

Of the wide range of material we have used in the teaching of cytology we have found the following particularly useful both in elementary and more advanced classes.

Male Locusts and Grasshoppers

These are undoubtedly the best types for demonstrating chiasmata at diplotene. Chiasmata and relational coils can be easily distinguished as, often, can the different types of double chiasmata. This material is equal to any and better than most for a detailed study of all the stages of meiosis. Grasshoppers like *Chorthippus* (2n = 16+ XX/XO) have both acro- and metacentric chromosomes in the complement, all of which can be individually distinguished. This makes them useful for demonstrating the difference between the anaphase of first and second divisions. The material shows also the chromomeric organisation of the early prophase chromosomes, the precocious condensation of chromosome ends, an XX/XO sex chromosome mechanism and the positive and negative heteropycnosis of the sex chromosomes at different stages of meiosis. The chromosomes of the desert locust, *Schistocerca gregaria* (2n = 23), are similar but all of them are acrocentric. Thus, *Chorthippus* and *Schistocerca* have the same number of chromosome arms. Meiosis occurs all the year round in laboratory-bred locusts and from July to September in adult grasshoppers. Mitosis can be obtained from the testes of late nymphal instars.

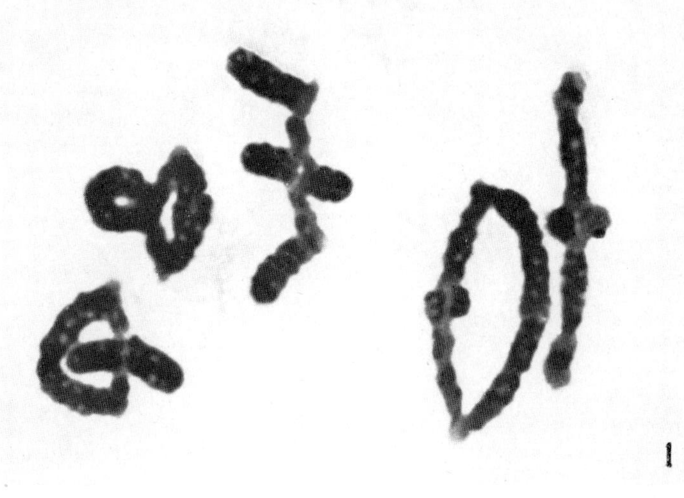

FIG. 1. *Paeonia lutea* × *delavayii* (2n = 10). Random chiasmata. Carnoy, aceto-carmine, × 1875.

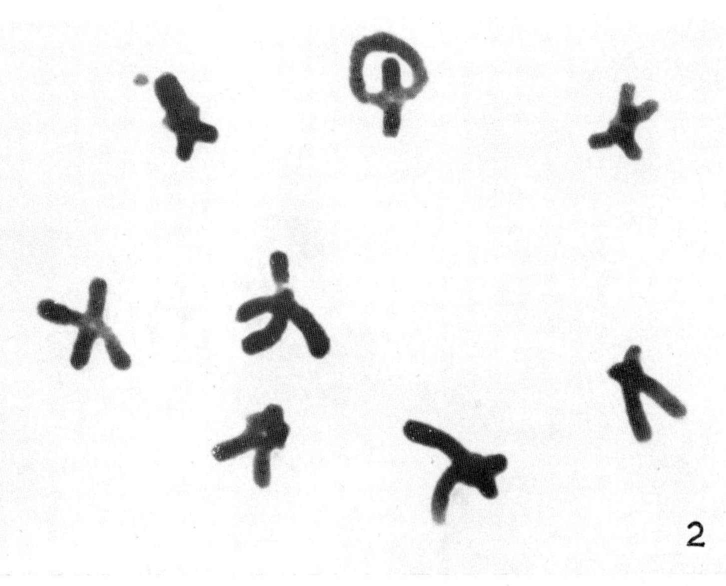

FIG. 2. *Allium fistulosum* (2n = 16). Only one of the chiasmata is terminal (ring bivalent at 12 o'clock). All others are localised near the centromeres. Benda, Feulgen, × 1500.

Preps and photos by K. R. Lewis and B. John

FIGS. 1 and 2. Metaphase I in PMC of Angiosperms showing different distributions of chiasmata (and crossing over), one, two or three per-bivalent.

Paeonia

Chiasmata are easily analysable at metaphase in this material where all the chromosomes are individually distinguishable in the diploid species (2n = 10). Many species hybrids, e.g. *P. lactiflora* × *tenuifolia* and *lutea* × *delavayii*, have an almost normal meiosis with only occasional failure of pairing. The chiasma frequency is usually low in all types and the tetraploids have a low frequency of multivalents; they also have bridges and fragments. Meiosis about April. Many anthers per flower giving a range of stages (Fig. 1).

Tradescantia

Very large chromosomes showing major coils at meiosis. Diploid, triploid and tetraploid material very easy to handle. Many clones have B chromosomes.

Male Newts

Easily analysed distally localised chiasmata at metaphase. Meiosis July–August.

Allium fistulosum

In contrast with *A. cepa*, most of the chiasmata in this species are localised near the centromere. It is, of course, much cheaper than *Fritillaria meleagris* which has the same centric localization. Meiosis in May (Fig. 2).

Humulus lupulus

This species has an XX/XY sex chromosome mechanism (cf. *H. japonicus*) where the larger X and smaller Y are easily distinguished. It is easier to handle than Melandrium where the Y is larger than the X which is not usual. Meiosis in late summer.

Rumex hastatulus

The flowers and anthers of this dioecious American species are rather small but the XY_1Y_2 multiple in the 'male' is very clear. There are only three pairs of autosomes which can be individually distinguished. Meiosis in May.

Narcissus 'Geranium'

That the onset of first anaphase depends on the lapse of chromatid attraction is beautifully illustrated by this numerical-cum-structural hybrid of N. poeticus (2n = 14) and N. tazetta (2n = 20). About three unequal bivalents are formed, the attraction between chromatids does not lapse and a restitution nucleus is formed. This enters second division with the major coils undone so that the chiasmata, which still persist, are very clear. Second anaphase also may be suppressed giving double non-reduction. It also illustrates segregational sterility for only diploid or tetraploid grains undergo mitosis. Meiosis in September.

MATERIAL FOR PRACTICAL CYTOLOGY

ANN P. WYLIE

Department of Botany, University of Otago, Dunedin, N.Z.

Polytene chromosomes in the black-fly, Eusimulium aureum

The giant chromosomes in larval salivary glands are bigger than those of *Drosophila. E. aureum* has the low chromosome number of $2n = 4$, and the two polytene chromosmes can readily be distinguished by length and centromere position. Balbiani rings are prominent in this species, and many individuals are heterozygous for one or more inversions. The distribution and frequency of inversions in Canadian and Eurasian populations have been intensively studied by Rothfels and Dunbar, who give detailed maps of the chromosomes.

The chief disadvantage of this species, compared with *Drosophila,* is that the material has to be collected from the wild, and is only available for a fairly restricted period in early summer, when the temperature of the water in the upland streams frequented by the larvae is not too high. The larvae occur on the undersides of stones in swiftly flowing streams in, for example, the Lake District and the Scottish Highlands. However the larvae can be fixed whole in acetic-alcohol and stored for long periods satisfactorily. The larvae are hydrolysed and stained with the Feulgen reagent. The glands, which are very large compared with the size of the larva, are then dissected out in 45 per cent. acetic acid. Orcein can be used to intensify the stain, should this be necessary. Permanent preparations are made by the 'dry ice' method of Conger and Fairchild.

Meiosis in anthers of paeony varieties

All the single-flowered varieties examined by me have proved to be diploids. They have many advantages as class material.

(a) They have a low chromosome number ($n = 5$) and the chromosomes are so large that a good deal can be seen with only a 4 mm. objective.

(b) One bud has a phenomenal number of anthers, and a very high proportion in a bud of the right age show stages of meiosis. Collecting material for large classes is no problem.

(c) Students are virtually certain to obtain a complete series of stages as one anther often shows a very wide range (probably a consequence of the hybrid nature of the material). It is common to find all stages from late prophase I to telophase II in the one preparation.

(d) Meiosis is not restricted to any particular time of day and no ill-effects have been detected from keeping buds for several hours (in a damp plastic bag) between being collected and being fixed.

(e) Anthers fixed in acetic-alcohol can be stored indefinitely in 70 per cent. alcohol in a refrigerator without deterioration. In fact, freshly-fixed material is unsatisfactory, as pollen-mother cells are still very fragile and easily damaged in squashing if preparations are made 2–3 days after fixation, and the intensity of staining is also poor. A satisfactory preparation is obtained if whole anthers are first hydrolysed and stained with the Feulgen reagent; a small piece of one anther is then squashed in aceto-carmine containing a good deal of iron. The covered preparation is then heated over a steam-bath for 10 minutes or longer. This clears the cytoplasm and flattens the cells. Permanent preparations are made by the standard method (Darlington and La Cour, 1963).

(f) Most bivalents have only 1 or 2 chiasmata and are easily analysed. The material is suitable for quantitative work by a class on chiasma frequencies. Terminalisation is not pronounced and it is easier to convince the sceptical student about chiasma structure than it is, for example, in *Tradescantia*.

(g) In the varieties examined by me, meiosis is sufficiently regular not to disturb the average student. However the enterprising student will find metaphase I cells with 2 (rarely 4) univalents and anaphase I and II cells with bridges and fragments; inversion heterozygosity is evidently rather common. The varieties differ significantly in frequency of aberrant cells.

The principal disadvantage of *Paeonia* as class material for meiosis is that the stages of prophase I up to late diplotene are not good with the method of preparation outlined above. The lack of deterioration on storage compensates for the material only being available for a restricted period of the year. Buds of the right age are quite large, 15–18 mm. in diameter, but still tightly enclosed by the very sticky sepals. The middle of May is the right time in an average year for plants growing in the open at Jodrell Bank, Manchester.

Chromosome morphology in Ipheion uniflorum (*syn.* Brodiaea Triteleia)

This attractive member of the Liliaceae (tribe Allieae) provides good material for demonstrating chromosomes with almost terminal centromeres. One of the six pairs of chromosomes has a submedian centromere, but in the remaining five pairs the centromere is very close indeed to one end of the chromosome, the second arm being a small, just visible knob. One of these pairs has conspicuous satellites.

Root-tip chromosomes respond well to colchicine pretreatment, but the morphology of the chromosomes at metaphase is also clear in untreated roots. Fixation in acetic-alcohol for 5–10 minutes is satisfactory. Considerable care is needed in bleaching roots fixed in 2BD, as they seem more liable to disintegrate in the process than do most roots.

Bulbs can be obtained commercially without difficulty and the plants grow all right in pots in a cold frame. Roots are ready for collecting in early autumn, when I have found extraordinarily high mitotic rates, even in spindly and outwardly unpromising roots.

Cytological methods in plant anatomy

Squash preparations of pieces of immature stem show well the nuclear changes involved in the differentiation of xylem and phloem, about which ordinary sections are most uninformative.

The first macroscopically visible internode of *Vicia faba* seedlings is suitable. This is cut into strips, which are fixed in acetic-alcohol. After hydrolysis and Feulgen staining, the pieces are dissected in 45 per cent. acetic acid under a low-power binocular. The vascular strands can be segregated from the cortex and pith quite easily, and divided into small pieces before covering. The amount of pressure needed to complete the squash is not great; it is important not to over-squash, as the sieve-tubes have rather plastic walls and are easily deformed. The cytoplasm is conveniently stained during dehydration by a few seconds in a dilute solution of Light Green in 95 per cent. alcohol. Preparations can be made permanent by either the 'dry ice' method or the standard one (Darlington and La Cour, 1963).

The enormous, highly endopolyploid nuclei in immature xylem vessel segments are very obvious, and stages in their disintegration at the end of maturation of the cells can be found. The differential cell divisions of a sieve-cell mother cell, to give the future sieve-tube element and companion cell, is easily studied and also all stages in the subsequent divergence of these two cell types. In *Vicia*, as in many Leguminosae, the mature sieve-tube element has a dense cytoplasmic inclusion, the slime-body, which appears during maturation, and frequently a second densely-staining body, probably a virus inclusion. Useful papers on the cytology of xylem and phloem are those of A. Resch (1954–61).

General points

(a) Growing *Vicia faba* for root-tips. Soak the seeds in water for 24–36 hours, until the testa has cracked at the hilum. Remove the testas; this ensures more uniform germination. Plant the beans in seed-pans containing damp vermiculite. When the radicle is $1\frac{1}{2}$–2 cm. long (about 48 hr., depending on the temperature), excise the apical 2–3 mm. and replant. Excision of the main apex stimulates the outgrowth of lateral roots, of which there will be a good uniform crop in 5–7 days. Keep the vermiculite damp, but not water-logged. Two pans (about 10 beans per pan) are plenty for a class of 50.

(b) Onions for root-tips. These are unsatisfactory in the autumn term because of dormancy. However, pickling varieties (e.g. 'Silver Skin') are all right in October. Dormancy in other varieties can be overcome by 4–6 weeks' storage in a refrigerator.

(c) Inflorescences of *Allium ursinum* are very good for meiosis (a recommendation from A. D. Bradshaw). I find that fixation in Newcomer's fluid (6 isopropyl alcohol, 3 propionic acid, 1 petroleum ether, 1 acetone, 1 dioxane) is superior to acetic-alcohol and its modifications. Material is still perfectly all right after two years in fixative in a refrigerator. It must be washed well before preparations are made (running water for at least one hour).

(d) I can verify the general superiority of *propionic-orcein* recommended by the Plant Breeding Institute, Cambridge, over aceto-orcein. The staining is more intense, and the solution can be kept at room temperature for several months without the necessity for filtering. It gives good staining of *Allium ursinum* pollen-mother-cells, and *Drosophila* salivary gland chromosomes, as well as of mitotic chromosomes of a variety of plant species. The stock solution is 2 per cent. in 45 per cent. propionic acid. Bring to the boil, cool, and filter. Dilute with an equal volume of 45 per cent. propionic acid for use.

(e) Material which is available only for a restricted season may be stored indefinitely for subsequent class use in 45 per cent. acetic acid in a deep-freeze unit, after it has been stained by the Feulgen method, prior to squashing (see Ford and Hamerton in *Stain Technology*, 1956, **31**, 297). This gives results far superior to the usual storage of fixed, unstained material in 70 per cent. alcohol at 0–2°C.

(f) I can recommend the 'England Finder' (marketed by Graticules Ltd., price 63*s*.) for locating individual cells under strange microscopes when putting up demonstrations.

REFERENCES

DARLINGTON, C. D., AND LA COUR, L. F. 1963. Handling of Chromosomes. Allen & Unwin, London.

RESCH, A. 1954–61.

Planta, 1954, **44**, 75–98
1955, **45**, 307–324
1958, **52**, 121–143
1959, **52**, 467–489
Z. Bot. 1961, **49**, 82–95

ROTHFELS, K. H., AND DUNBAR, R. W. 1958 and 1959. *Canad. J. Zool.* 36, 37.

PERIPHERAL BLOOD CULTURES OF HUMAN CHROMOSOMES

C. H. OCKEY
Paterson Laboratories, Christie Hospital and Holt Radium Institute, Withington, Manchester

The introduction of improved techniques into the study of human chromosomes has led to the discovery of abnormal complements associated with certain clinical syndromes. Human chromosomes have therefore become of interest to the medical profession and new chromosome conditions appear regularly in the medical journals. Cultures of peripheral blood offer suitable student material for human chromosome study. Very little extra equipment than that present in the general biological laboratory is demanded. The blood can usually be obtained under medical supervision from a member of the class with little inconvenience.

Culture technique

The method described is based on modifications of Moorehead *et al.* (1960) by the MRC Group for clinical effects of radiation, Edinburgh, and by ourselves.

A sterile 3½ in. needle and short air-vent needle, the latter fitted with a short rubber tube plugged with sterile cotton wool, are introduced through the washer of a sterile 1 oz. McCartney or Universal bottle (the washer having been coated with collodion). 1 ml. of 0·4 per cent. heparin in distilled water is introduced and the whole placed in ice-water. Immediately before the blood is introduced, 0·4 ml. of diluted Phytohemagglutinin 'P' is added (1 ml. from the 5 ml. vial is diluted with 9 ml. of Phytohemagglutinin buffer before use). A sterile 20 ml. all glass syringe, chilled and washed out with sterile heparin, is used to obtain the blood. The 20 ml. of blood is immediately injected into the bottle through the 3½ in. needle and the bottle returned to the ice-water.

After 30–50 mins., the blood is centrifuged for 5 mins. at 350–400 rpm. The leukocyte and plasma layer overlying the agglutinated red cells is removed carefully with a sterile, all glass, 10 ml. syringe and introduced into a second sterile bottle. An alternative method of agglutination of the red cells without centrifuging is to use 1 ml. of 6 per cent. Dextran; the Phytohemagglutinin is then added to the plasma (0·02 ml. per 1 ml.).

It is important not to introduce red cells with the plasma or deterioration of the cultures may result. After mixing, a white cell count is carried out (4,000–8,000 cells per mm.^{3}). This is then diluted with tissue culture medium '199' (pH 7·4–7·6) until the count lies between 1–2,000 cells per mm.^{3} (usually 2 ml. plasma to 8 ml. '199'). The pH of the commercial '199' requires adjustment with sterile 4 per cent. Sod. Bicarb. before use.

After mixing, 7–8 ml. aliquot parts of the culture are set up in sterile bottles and incubated at 37°C.

The Phytohemagglutinin induces mitosis in the immature blast cells 2–3 days after incubation. At this stage cells are harvested. The quality of the preparation is dependent on the pH of the culture at this time. The pH usually drops from 7·4 to 6·8 in 2–3 days. If however it drops below 6·6 few divisions occur and the chromosomes appear 'fuzzy' and do not spread. A high white cell count in the culture can cause this rapid drop. The pH also drops rapidly with leukaemic cells, these are harvested generally at 1–2 days. As approximately 4 cultures are obtained from 20 ml. blood it is advisable to harvest at intervals from 2 to 3 days. Three hours before harvest 0·1 ml. of 0·04 per cent. colcemid (CIBA) at 37°C. is added per 1 ml. of culture.

Preparation of white cells

After colcemid treatment, the culture is well mixed, pipetted into a 10 ml. centrifuge tube and spun at 1,000 rpm. for 5 mins. The supernatant is removed, leaving a drop to resuspend the cells. Freshly prepared 0·95 per cent. sod. citrate is then added (5 ml.) to swell the cells and the tube incubated at 37°C. for a further 20–30 mins. The tube is then spun as before and the supernant removed. The cells are thoroughly mixed in one drop of this supernant. 5 ml. of fresh fixative (3 : 1, abs. alc.: gl. acet. acid) are added drop by drop, shaking the suspension after each drop to prevent the cells from clotting. After half an hour fixation at 0–5°C., cells are spun down and the old fixative replaced with fresh (1–1·5 ml.); cells are resuspended, stored at 0–4°C. or slides made at once. Replace fixative with fresh 3:1 methyl alc.: gl. acet. acid, repeat, then resuspend in 1 ml. of 1:3 methyl alc.: gl. acet. acid for 15 mins.

Preparation and staining of slides

The air-drying method (Rothfels and Siminovitch, 1958) is preferred to the squash technique as the cell membranes are not ruptured and chromosomes expelled.

Grease-free, clean slides are immersed in a beaker of ice and distilled water for several minutes. A slide is removed, excess water shaken off and 2 drops of the cell suspension dropped on to the centre of the cold, moist slide. After about 10 seconds of spreading, the excess water and fixative is shaken off and the slide passed horizontally, for about 30 seconds, through a spirit flame to dry. Slides can be stored or stained immediately in acetic orcein for 1–2 hours. Permanent preparations are made by passing through (a) 45 per cent. acetic acid—5 dips. (b) Abs. Alc.—1 min. (c) Abs. Alc.—1 min. (d) 'Euparal' essence—2 mins. (e) 'Euparal' and coverslip. One culture will yield material for approximately 15 slides.

Equipment notes

3½ × 18 g. Milward's exploring and 'Viking' No. 1 short bevel record needles available from: Shrimpton and Fletcher Ltd., Redditch.

'McCartney' or 'Universal' bottles and Difco Phytohemagglutinin P and buffer: Baird and Tatlock.

Tissue culture medium '199': Glaxo Laboratories, Greenford, Mx.

Heparin (crystalline): L. Lights and Co. Ltd., Colnbrook.

Diversey pyroneg (for washing syringes, needles, bottles): Deosan Ltd., 42–46 Weymouth St., W.1. Cleaning is followed by 24 hours' running tap water, and 24 hours' distilled water.

Natural Orcein: G. T. Gurr Ltd., S.W.6. 1–1½ gm. are boiled in 60 ml. acet. acid for 10 minutes, cooled and 40 ml. distilled water added, (filtered before use).

Sterilisation of equipment: Autoclave 15 mins. at 15 lb.

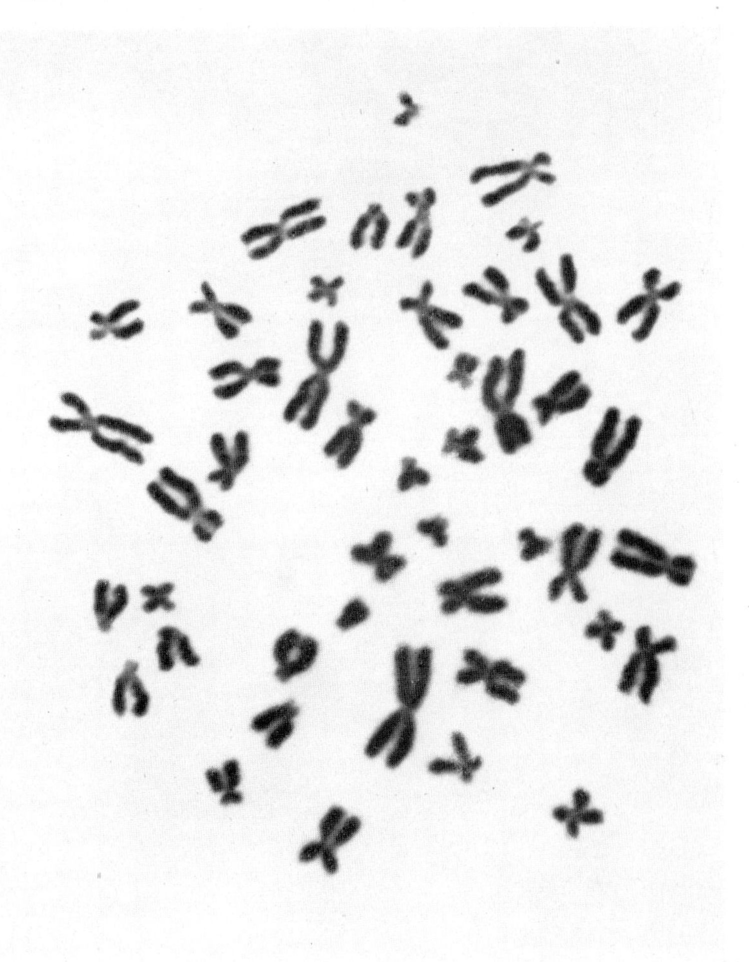

FIG. 1. Chromosomes of a normal male from a 3-day peripheral blood culture. Magnification × 2,500.

Chromosome analysis

The human chromosomes (Fig. 1) are classified according to the 'Denver' classification (1960). The six large acrocentrics (13–15) can be used as standards for analysing the remainder. They are satellited and often show nucleolar association with each other and with Nos. 21, 22. No. 15 has the smallest satellite. Of the 4 or 5 small acrocentrics, 21, 22, and Y, the Y has a larger short arm and its chromatids tend to lie parallel.

The small metacentrics 19 and 20 are rarely distinguishable, 19 is slightly sub-median and larger. No. 16 is a larger metacentric, but can be confused with 17.

No. 18 is smaller than 17 and has a more median centromere.

No. 1 is the largest metacentric, but one arm may appear longer due to stretching and thus resemble No. 2 which is sub-median. No. 3 is a metacentric smaller than No. 1, Nos. 4, 5 are difficult to separate except by length in good cells. No. 6 forms the largest pair of the sub-median group X, 7–12, it may on occasions resemble a No. 3.

The only method of further analysing the X-7–12 group is by pairing up enlarged cut-out photographs (Muldal and Ockey, 1961a). The presence of an extra X chromosome in this group is often confirmed by the study of sex chromatin.

REFERENCES

MOOREHEAD, P. S., NOWELL, P. C., MELLMAN, W. J., BATTIPS, D. M., AND HUNGERFORD, D. A. 1960. Chromosome preparations of leukocytes cultured from human peripheral blood. *Exp. Cell. Res.*, **20,** 613–616.

MULDAL, S. AND OCKEY, C. H. 1961a. The Denver classification and Group III. *Lancet*, **2,** 462–463.

MULDAL, S. AND OCKEY, C. H. 1961b. Muscular Dystrophy and deletion of the Y Chromosome. *Lancet*, **2,** 601.

ROTHFELS, K. H. AND SIMINOVITCH, L. 1958. An air-drying technique for flattening chromosomes. *Stain Tech.*, **33,** 73–77.

VARIOUS AUTHORS. 1960. A proposed standard system of nomenclature of human mitotic chromosomes. *Lancet*, **1,** 1063–1065.

BLOOD GROUPS IN THE TEACHING OF GENETICS

A. E. Mourant

M.R.C. Blood Group Reference Laboratory, Lister Institute, Chelsea Bridge Road, London, S.W.1

As I have myself little experience of the formal teaching of genetics I do not feel competent to present you with a syllabus for a set course of training. It will I think be more valuable to you if I suggest a number of topics within the blood group field which appear to me to illustrate important principles, in the hope that we may in subsequent discussion see how they can be fitted into a syllabus.

I assume that basic genetical principles will already have been mastered, and that the class will use themselves as experimental animals in order to illustrate these principles, at the same time studying the serological and genetical theory of the blood groups.

For the benefit of those teachers who are not medically trained I must first say something about the dangers of bleeding students, by which I mean the dangers of infection. Some students may indeed faint, even before they are touched with a needle, and may need first aid. The dangers of infection are however much greater, though less obvious.

For some purposes 1 ml. or more of blood may be needed; in this case it is best that a medical man should perform venepunctures, and he will then take responsibility for sterility. I might point out, however, that a fresh heat-sterilised syringe and needle must be used for each person. One possibility is to use cheap plastic disposable syringes, supplied ready sterilised by the manufacturers.

If only a few drops of blood are needed, this may be obtained by sharply stabbing a finger or the lobe of the ear with a triangular surgical needle. Students may bleed one another or, in the case of the finger, themselves. The bleeder's hands should be scrupulously clean; the site of the prick should be cleansed by rubbing firmly several times with a swab moistened with spirit, the main object being to remove an invisible layer of infected grease. A separate sterile needle should be used for each prick. Needles should be heat sterilised by a bacteriologically approved method (baking at 160°C. or autoclaving at 120°C.): waving in a flame may render the tip red hot yet leave living organisms farther up. While there is in every case a slight danger of bacterial infection of the wound, the chief danger is the transfer from the blood of the occasional symptomless carrier to other people of the virus of serum hepatitis, a disease with a long incubation period and quite an appreciable mortality. This virus is highly resistant to chemical disinfectants such as alcohol, and epidemics are notoriously liable to occur in clinics where numerous people receive skin pricks.

The reagents used in blood grouping mostly consist of human or animal serum, and bacteria grow readily in them. This may give rise either to false positive or false negative results in tests. Sera must therefore be kept in a refrigerator. Unless the label indicates otherwise, the main supplies should be kept frozen solid, but not frozen and thawed repeatedly as this also weakens them. Because of the changes which may take place in reagents it is important that in all series of tests of unknown red cells there should be included control positive and negative tests, i.e. tests upon known cells which should give positive and negative results respectively with the reagent in use.

The ABO blood groups

Most of the testing reagents used in blood group work consist of serum obtained from human beings, and the only such sera which are available in sufficient quantities for use in class work, other than for a limited number of medical students, are those for detecting the antigens of the ABO genetical system.

I do not propose to describe now the well known genetics of this system. Nor shall I describe the methods of testing, which are given in a number of standard works, except to say that if class work is to become at all extensive, it is essential that methods should be used which are economical of serum. It is best that the teacher should receive some instruction in the performance of the test in small (precipitin) tubes at a transfusion centre or hospital laboratory. Lecture experiments, done only by the teacher, can sometimes legitimately be such as to use rather more serum.

How then can we use these tests to illustrate genetical principles. Ideally, from the genetical point of view, each student should be encouraged to obtain fresh finger or ear prick samples of members of his family, as well as from himself, and to test all of these. The combined results from a whole class and their relatives should give a convincing demonstration of the genetics of the system. The possibility must be faced of bringing to light an occasional case of extra-marital paternity, which may have serious social consequences. One way of forestalling this would be to distribute the specimens for testing in the first instance identified solely by numbers. Alternatively, the class, if a large one, can be treated as a sample of a population in breeding equilibrium and used to illustrate the Hardy-Weinberg law.

The secretion of the antigens of the ABO system

Tests on saliva for the secretion of blood group antigens involve the use of the same two sera as for ABO blood grouping, together with a third reagent, anti-H, for the basic antigen present in group O cells as well as (in reduced amount) in those of other groups. Each student can use his own cells when testing his saliva, but it may be preferable to use standard suspensions of cells of groups O, A and B obtained by venepuncture. The best anti-H reagent for these tests is an extract of the seeds of common gorse, *Ulex europaeus*. The secretor phenomenon may be used as an example of a two-gene system with simple dominance, and also as an instance of epistasis, since, for instance, if A substance is to be secreted the subject must have both an A gene and a secretor gene.

A theoretical treatment of the genetics of the Lewis system may be brought in at this stage, but the reagents are too scarce and the tests too difficult for them to be carried out by a class.

The Rh system

It is essential that the genetical theory of the Rh blood groups should be included in any course, but the inclusion of

Rh tests in the practical course raises difficulties, especially regarding the supply of serum.

Nearly all the reagents required are derived from women who have been immunised by pregnancy, and only very exceptionally will any other sera than anti-D be available for class use, and even the latter is in permanent short supply. It is best that anyone contemplating using it should make a personal approach to the Director of the Regional Blood Transfusion Centre serving his area. Anti-D of the variety which agglutinates untreated D-positive cells suspended in saline is very scarce indeed. Only a serum of the 'incomplete' variety is likely to be available, and this needs to be used by one of a number of special methods. For class use probably the best way will be to prepare a mixture of antiserum and papain solution (Löw's technique: see Medical Research Council, 1958, Mem. No. 36, p. 27) and to check immediately before the class assembles that it gives reliable positive and negative results. This method has the further advantage that it works quite well with sera which are on the borderline of usability for clinical work, and which are therefore more easily spared by transfusion centres than are stronger sera.

It is desirable that theoretical teaching should include an account of haemolytic disease of the newborn, and of the association of certain other diseases, such as gastric carcinoma and duodenal ulcer, with particular blood groups or with the non-secretor type. The relationship between thyroid diseases and P.T.C. tasting might conveniently be mentioned here. The linkage of certain congenital abnormalities, and of secretion, with particular blood group systems should also be mentioned. It is especially important to make clear the distinction between association and linkage, as I have found much confusion between the two concepts in the minds of persons proposing to carry out clinical research.

It is unlikely that the reagents for any of the blood group systems other than ABO and Rh will be available in quantities sufficient for class use; and indeed teachers, unless they have had training in a blood grouping laboratory, may find that they themselves have serious difficulties in getting consistent results when demonstrating the reactions of such sera.

Apart from American pharmaceutical firms which charge

very high prices indeed (of the order of pounds per ml.) the only source of testing sera in this country is the National Blood Transfusion Service. My own laboratory can supply small quantities of anti-A and anti-B, and I shall be glad to arrange this as far as I can for serious genetical courses, but not for courses in such subjects as hygiene or human biology without serious genetical content.

A convenient source of some reagents for those prepared to take the trouble to prepare them, is the seeds or other parts of certain plants. Those who have access to botanical departments may think it worth while not only to prepare the reagents but to grow the plants. Most of the reagents (anti-A, anti-A_1, anti-H, anti-N) are prepared from leguminous seeds. It appears that the only reliable plant anti-B comes from a fungus (*Marasmius oreades*) found mainly in Scotland and northern Europe; anti-M is said to be obtainable from *Iberis amara* but so far only extracts have been on the market and the commercially available seeds (and others of the same genus) have not given any positive results at all in my own laboratory. I believe others have been more fortunate. Botanical geneticists may like to experiment with species and strains of this genus as well as of those producing other agglutins. My own laboratory can supply small quantities of seeds of some of the species mentioned in the literature to those who wish to grow them.

REFERENCES

Theory (elementary)

LAWLER, S. D. AND LAWLER, L. J. 1957. *Human blood groups and inheritance*, 2nd Ed. Heinemann, London.

Theory (advanced)

RACE, R. R. AND SANGER, R. 1962. *Blood groups in man*, 4th Ed. Blackwell, Oxford.

Laboratory methods (elementary)

MEDICAL RESEARCH COUNCIL. 1958. *The determination of the ABO and Rh (D) blood groups for transfusion*. Medical Research Council Memorandum No. 36. H.M. Stationery Office, London.

Laboratory methods (intermediate)

BOORMAN, E. AND DODD, B. E. 1962. *An introduction to blood group serology*, 2nd Ed. Churchill, London.

DUNSFORD, I. AND BOWLEY, C. C. 1955. *Techniques in blood grouping.* Oliver and Boyd, Edinburgh.

Laboratory methods (advanced)

STRATTON, F. AND RENTON, P. H. 1958. *Practical blood grouping.* Blackwell, Oxford.

Population data

MOURANT, A. E. 1954. The distribution of the human blood groups. Blackwell, Oxford.

MOURANT, A. E., KOPEC, A. C. AND DOMANIEWSKA-SOBCZAK, K. 1958. The ABO blood groups: comprehensive tables and maps of world distribution. Blackwell, Oxford.

Recent advances

GOLDSMITH, K. L. G. (Scientific Editor). 1959. Blood groups. *Brit. med. Bull.* **15,** 89–174.

The Proceedings of the Second International Conference of Human Genetics, Rome, 1961, will contain papers, with new population data, on human blood groups and other polymorphisms.

THE GENETIC GARDEN

C. D. Darlington
Botany School, Oxford

As genetics shapes itself to be the framework of the life sciences —and escapes from its mathematical birth pangs—more and more the teacher of genetics will become aware of the immense task that faces him. He will want to get living plants and animals to do his teaching for him. Most ways of doing this are costly in space, labour and equipment.

One problem is that of explaining chromosomes. In future projection, directly or by television, will be found indispensable. Models of chromosomes which now have to be made by hand will be mass produced. They should show normal meiosis with normal crossing over and disjunction and also with crossing over and disjunction in inversion and interchange hybrids. The inability to teach this part of the subject hamstrings genetics in many Universities. Another problem is how to maintain stocks and collections. What materials can be kept readily available for teaching and research? Human beings provide one answer: their genes and chromosomes are always handy in blood, spittle and urine.

A garden of genetic variation gives us another answer: it merely requires co-operation in exchanging plants. And it can almost be made to speak for itself.

The necessary plants which I have collected in a Genetic Garden are chiefly hardy perennials. This saves supervision and ensures permanence and standardisation. They fall under four heads

Origins of Species

The main lesson should be to show the paradox of variation. On the one hand the chromosomes change without externals in

Allium or *Trillium* (where every individual can be recognised by its mitotic chromosomes). Similarly in *Campanula persicifolia* which can be used to show interchange hybridity in all wild stocks, telocentrics in some, and a tetraploid in the garden form, Telham Beauty. On the other hand externals change without the chromosomes in maize. Here artificial selection has turned *Euchlaena* into a new genus *Zea* without grossly altering the chromosomes. Among the ornamental gourds the turban gourds are notable in showing (summer and winter) how the fundamental distinction between the inferior and superior ovary can be overridden by artificial selection.

Species crosses reveal another side of the problem and are most conveniently found in ornamental genera. In *Narcissus* we have the clonal variety *Geranium* illustrating the origin of new basic numbers of chromosomes: *N. poeticus* × *tazetta* ($10 + 7 = 17$). In *Rosa* we have a series of diploid and polyploid forms, both fertile and sterile, which illustrate the origin of garden roses of the past and also continuing today. The well known blackberry-raspberry hybrids (Loganberry and Veitchberry) are nearly true breeding polyploids. Another species cross, the John Innes berry, shows tetraploid segregation for spines in the seedlings.

Breeding Systems

Incompatibility, heterostyly and dioecism are the polymorphisms which fix these systems and must be illustrated. For comparison we have unfixed or floating polymorphisms such as that for cyanogenesis in *Trifolium*. *Forsythia* as well as *Lythrum* and *Primula* species illustrate heterostyly. For incompatibility a striking example (due to Karl Sax) is *Syringa laciniata* which if grown as a single plant will always hybridise with any neighbouring lilac (*S. vulgaris*) to give the sterile cross known as *S. persica*.

On the chromosome side special systems can be represented by complex hybridity (as with rings in *Rhoeo*) and methods of generating diversity (as with iso-chromosomes in *Nicandra*). Triploids can be shown giving different degrees of sexual sterility (in apples and pears) or avoiding it by apomixis in *Taraxacum*. *Ranunculus auricomus* in 2x and 4x apomictic forms is notable for its immense range of performance in the endosperm varying from 4x to 16x. It is also pseudogamous (Rutishauser 1960).

Mutation, Variegation and Graft-Hybrids

All degrees of instability of genes, chromosomes, plastids and plasmagenes can be illustrated and related to one another by observation during development as well as by breeding. Mutable genes affecting flower colour are shown by flaked forms of *Antirrhinum, Dahlia* and *Zinnia.* These should be contrasted with breaking of flower colour due to virus in *Tulipa.* The two old flaked roses, York and Lancaster and *Rosa Mundi,* show gene mutation in the growing point contrasted with that confined to the flowers.

Mutable plastids show the same contrast. In the *iojap* maize, and in a clone of *Ballota nigra* in the Oxford Botanic Garden, plastids mutate in all parts of the plant including the growing point. But in the flaked variety of the Norway Maple *Acer platanoides Leopoldii* mutation is confined to the leaves.

Two other plants well known in horticulture are worth noting in this connection. *A. platanoides Drummondii* is the best white-over-green chimaera among trees. And *Spiraea bumalda* shows a spurious appearance of chlorophyll mutation since it looks green but gives white shoots: in fact it is a chimaera with an invisible white epidermis and the mutation is due to sorting out of tissues in the chimaera.

The doyen of graft hybrids, *Cytisus Adami,* can be kept to show both permanence and instability. As a footnote to the theories of Lysenko the contrast between a mutable graft hybrid and a fixed sexual hybrid may be represented by *Crataego-Mespilus* in which both kinds of hybrid are available: *C.M. dardari* is a graft hybrid, *C.M. grandiflora* a sexual hybrid.

Grafting itself should not be forgotten as a genetic technique in a collateral sense. A collection of seven bud sports of the triploid Blenheim Orange apple can in this way be mounted economically on one stock.

The whole range of green-over-white and white-over-green chimaeras can be represented in the cultivated ivies and hollies. For breeding purposes, crossing white and green cells and producing mixed plastid eggs which sort out in development,

Pelargonium and privet (*Ligustrum*) are to be recommended. For physiological experiments with variegation *Lonicera japonica* and *Populus canadensis* 'Aurora' are most convenient. *Pelargonium* chimaeras show a wide genetic range of types differing in the plastid or the nuclear control of their physiological properties.

Most rose growers will provide thirty climbing varieties of Hybrid Tea roses together with the bush varieties from which they have arisen as bud sports. These provide examples of presumed cytoplasmic mutations which can be conveniently contrasted with giant polyploids in *Antirrhinum* as well as in *Rosa*.

Multi-Purpose Plants

These are especially important in saving time and space. In *Melandrium* we can combine species crosses, sex chromosomes, and variegation due to chlorophyll deficiency. In *Tradescantia* with new aneuploids, asynaptics, telocentrics and interchange rings we can combine material for studying the structure of chromosomes, their experimental breakage, pollen grain genetics, species crossing and the production of unbalanced forms and new kinds of chromosomes. In *Allium* we can combine species crosses with the study of genotypic and structural control of chiasmata and of meiosis.

A garden with such genetic plants lets us see not just the results of evolution but how it is happening. And if one were to build one's genetics teaching round such a garden it would be genetics with its windows wide open on all sides.

REFERENCES

BRIERLEY, J. K. 1961. Some suggestions for the teaching of evolution in the field, garden and laboratory. *School Science Review*, **148**, 401–410.

DARLINGTON, C. D. 1963. *Chromosome Botany and the Origins of Cultivated Plants*. Allen and Unwin, London.

DARLINGTON, C. D. AND LA COUR, L. F. 1962. *The Handling of Chromosomes*. (4th Ed.) Allen and Unwin, London.

DARLINGTON, C. D. AND ROBINSON, G. W. 1957. *Oxford Botanic Garden: Guide*. Blackwell, Oxford.

DARLINGTON, C. D. AND SHAW, G. W. 1959. Parallel polymorphism in the heterochromatin of *Trillium* species. *Heredity*, **31**, 89–121.

RUTISHAUSER, A. 1960. Die Evolution pseudogamer Arten. *Ber. Schw. Bot. Ges.*, **70**, 113–125.

SAX, KARL. 1945. Lilac Species Hybrids. *J. Arnold Arboretum.* **26**, 79–84.

WATANABE, H. 1962. An X-ray induced strain of ring-of-12 in *Tradescantia paludosa. Nature*, 193–603.

GENETICS IN SCHOOLS

A. D. Bradshaw

Department of Agricultural Botany,
University College of North Wales, Bangor

To the average school child genetics is something to do with a monk named Mendel who lived in Switzerland in the middle of the last century and who was the author of two laws: chromosomes are things found inside the nucleus which undergo curious gyrations that are very difficult to remember but which have something to do with heredity; and evolution has nothing to do with either. This may seem an exaggeration. But a sample of scripts from a range of schools in answer to any questions on genetics set in an advanced level G.C.E.* examination will confirm this, despite the fact that it is an integral part of the biology syllabus of every examining board in the country.

The truth is that, at the present day genetics, not even well taught in all University departments, is hardly taught at all in some schools. Yet the situation is understandable. Although the rediscovery of Mendel's work is over sixty years away, genetics did not come to be the comprehensive subject that it is now, and the general biological significance of its findings did not become generally appreciated, until the period between the wars. By this time a large proportion of present day biology teachers had finished their University careers, without meeting any more of genetics than a passing reference to the work of Morgan or Bateson and Punnett. Many of them will have had little time or inclination to remedy the deficiency and until the recent awakening of popular interest in genetics will have rarely had genetics forced upon them in their everyday reading matter.

What then is needed? The general findings of genetics are now covered very adequately for schools by many textbooks

* General Certificate of Education.

from both genetical [2, 3, 8, 16, 17, 18, 24] and evolutionary [12, 13, 14, 20, 27, 28] points of view. But although the schoolboy has been taught genetics he usually scarcely appreciates the way in which the whole subject fits together, or the relevance of its findings to biology in general. How then, within the bounds of school time and facilities, can genetics be made intelligible, interesting, and significant? Before I try to offer positive suggestions let me first consider two general weaknesses of teaching biology which are, from my experience, particularly relevant to genetics.

Firstly, the *descriptive, historical* approach. The subject is presented in a manner which only memory can master, its contents separate and disconnected, ordered only historically. Such an approach is the complete antithesis of scientific teaching. Genetics more than many other aspects of biology, is logical and amenable to strict logical development. Yet the average schoolchild usually lacks any appreciation of the logic of genetics and the inter-relationship of its various findings even at the simplest level.

Secondly, the *biology can be learnt from textbooks* approach. The divorce of biology from the roots from which it has been developed, the disregard of practical experiment and observation, is often justified on the grounds of shortage of time or facilities. Practical classes take time and energy and indeed, appear so wasteful of these resources, especially in relation to an already packed time-table, that it may seem quite reasonable to cut them to a minimum. But biology dries up when practical work is eliminated, and its particular attraction to many children is removed. Genetics falls too easily to this approach, perhaps because it can be presented without practical work quite easily, but also because it is believed that practical exercises in genetics are difficult to carry out at school. However, neither argument is sufficient to justify the desiccation of genetics into a rather unpalatable sort of mathematics.

If these are the weaknesses how can they be obviated and how can genetics be taught in both a logical and practical manner? The following scheme is one solution. It may be too long, but this is to show what can be done: it can always be trimmed where necessary. The only essential facilities apart from those of a normal school laboratory are a corner where small animals can be kept, a small part of the school gardens where miscellaneous plants can be grown, and windows provided with shelves on which potted plants can be grown (or a greenhouse).

The individual has a development and an existence. It can be altered by the environment but it cannot be changed permanently.

$$\text{Genotype} \longrightarrow \text{Phenotype}$$
$$\nearrow$$
$$\text{Environment}$$

A clone of a vegetatively propagated plant such as 'Wandering Jew' (*Zebrina pendula*) can be split into pieces and grown on in a variety of different surroundings, in full light or half light, in wet or dry, outside or inside, for a term until the different plants are quite distinct. Then, at the beginning of the next term sample cuttings of them all can be taken and grown under one condition to see whether any modification has become permanent. The same can be done with a stock of *Drosophila.*

After such a study the underlying chromosomal mechanism which determines the fixity of the genotype must be considered[23]. The process of mitosis can most easily be seen by acetocarmine squashes of onion roots previously grown in jars of tap water. The use of Thomas's technique[10] of adding a drop of saturated ferric chloride solution to the acetic alcohol fixative makes the technique so easy that even a complete novice can succeed in an hour. The variation of chromosome number from species to species can be shown by giving some of the class a contrasting species such as *Crocus balansae* ($2n = 6$). If anyone wishes to experiment for himself in a free period the effect of 0·05 per cent. colchicine for 24 hours makes a good investigation. Animals present more difficulties but a slide of *Drosophila* salivary gland chromosomes is an excellent demonstration and is not difficult to make [10, 19]. The pattern of the chromosomes never fails to capture imagination. Wild *Drosophila* can easily be caught on rotten fruit, and the larvae cultivated on suitable media.

In contrast to the constancy of vegatative propagation is the variability provided by sexual reproduction. Such variability can never be overstressed since it emphasises the fundamental characteristics and significance of sexual reproduction. There are many vivid examples in plants. Perhaps the contrast of a potato variety and a set of seedlings raised from it is the easiest demonstration to make, but a more permanent demonstration is a row of seedlings grown from pips taken from a Cox's Orange Pippin apple.

The underlying mechanisms of chromosome behaviour in sexual reproduction now deserve examination. An excellent description has recently been published[23]. Meiosis can be introduced as a necessary complement to sexual fusion, a means by which the chromosome number is halved. With this approach its characteristics are readily appreciated. But then, in meiosis other things must be pointed out: that bivalents orientate and segregate independently of one another, demonstrable only where there are at least two pairs of dissimilar homologous chromosomes, and that chiasmata occur, the formation of which must be due to some sort of breakage and recombination.

As a result there are three levels at which variation is introduced into sexual reproduction, the levels of gametes, chromosomes, and parts of chromosomes. At the same time, despite all this, the basic continuity of the chromosomal material must not be forgotten. Other features can be left severely alone: yet how often are details of centrioles and nuclear membranes remembered and these points forgotten?

Meiosis is readily observed in both plants and animals using acetocarmine or acetic-orcein squashes[10]. The testes of grasshoppers caught in the summer is perhaps the best material, particularly because chiasma formation in early prophase is very clear and because a single preparation is likely to have all stages of meiosis on it. Preparation by an aceto-carmine squash following fixation using acetic alcohol and ferric chloride is not difficult. However, plant material is often easier to collect and the chromosomes are larger. Bluebell bulbs dug up in early February, wild garlic (*Allium ursinum*) inflorescences picked when 1–2 in. high in March, both provide excellent material in which there is a range of bud sizes, some of which should be undergoing meiosis. *Lilium regale* buds taken when about 2 cm. long and single peony buds about 16 mm. long are also excellent. Perhaps the most convenient plant is *Tradescantia paludosa* which flowers all the year round providing it is kept indoors in the winter: it is readily raised from cuttings. All are best squashed in acetic orcein: wild garlic does not deteriorate markedly over a year fixed in acetic alcohol and kept in a refrigerator.

Only now should the findings of Mendel and the problems of linkage and recombination be introduced, even if historically this is the wrong way round. For now the curious, at first sight inexplicable, laws of inheritance become immediately intelligible in terms of the basic pattern of meiosis. The most effective material for the demonstration of inheritance is unquestionably *Drosophila*[19]. Indeed, it is so vivid, particularly if the class made the crosses themselves, that the slight trouble involved is justified. Suitable stocks should be obtainable from biological supply houses quite cheaply obviating the need to keep the stock going all the year, and the medium is no more difficult to make than porridge. However, plants need not be disregarded, for although results take longer the characters are more easily observable, and the material can often be preserved on herbarium sheets. Good material is *Antirrhinum*[4] and the rayed form of groundsel *Senecio vulgaris* in which the heterozygote is recognisable and can be allowed to self automatically. In the South maize may be grown out of doors with some success, otherwise a greenhouse is necessary. It is unlikely that there are many schools where groundsel cannot be grown somewhere.

Many other cultivated plant species with pure breeding varieties of contrasting colour are possible such as stocks and sweet peas, as are some small animals such as mice and guinea pigs. Species which have small family size such as guinea pigs are not necessarily a disadvantage; they merely illustrate the problems of sampling errors more vividly. Each species presents its own problems but has its particular interest, and the choice can be left to individual interest, and to considerations of space and the other uses to which the material can be put (e.g. *Antirrhinum* for herbaceous borders and guinea pigs for dissection). However, what is certain is that some breeding experiments can and should be in progress in every school biology laboratory.

Some material of flower colours, e.g. *Antirrhinum*, can be used to illustrate the mechanism of gene action, if there is time and interest. Paper chromatograms work extremely well on the self-coloured flower pigments[4].

At this point most people might feel that they had dealt very fairly with the basic characteristics of the genetic mechanisms. Yet about nine tenths of all observable genetic characters, including those affecting most economic features of plants and animals, vary in a continuous and not a Mendelian manner and have patterns of inheritance not explicable in simple Mendelian terms. For this reason it is very worthwhile to introduce the concept that genes can have small additive effects. If the consequences of such an assumption are determined in a genetic system that is otherwise normal, the typical patterns of inheritance of continuously varying characters can be explained[11].

A survey of the basic genetic mechanism now completed, the ways in which variation between individuals originates must be considered. The primary cause is clearly the recombination provided by the normal processes of sexual reproduction, allowing innumerable different gene recombinations without loss or dilution. But the most fundamental cause is mutation. It is sometimes a difficult concept to believe, but a number of plants such as larkspur, dahlia and chrysanthemum provide varieties containing genes with sufficiently high mutation rates for mutation to be seen in the vegetative parts of the plant. Polyploidy can readily be demonstrated in 'giant' varieties of some garden species, e.g. *Antirrhinum* since these are autotetraploid, or by specimens of some of the more famous allopolyploid plants such as ordinary bread wheat, *Triticum aestivum*, or *Spartina*.

But a discussion of variation would be incomplete without a consideration of the effects of breeding systems. Evolution is unthinkable without cross fertilisation. This is elegantly illustrated by the variability to be found in rye (a normal outbreeding species) in comparison with wheat (an enforced inbreeder). It can be seen in fields or in a small plot of each. Rye perhaps provides the finest readily available example of the variability that is normally to be found in a species. In this discussion there will be a place for a consideration of the biological significance of sexual differentiation in animals and of all the various and remarkable forms of flowers in plants, and for the evolution of genetic systems as a whole[9].

Observations on variability lead on to considerations of

evolution, perhaps because such observations more than anything else have confirmed Darwin's ideas. There is a general tendency for schoolchildren to have a very woolly idea of the mechanism of evolution similar to their vagueness over the mechanism of inheritance. They have all heard of Lamarck and Darwin and are sure Lamarck was incorrect, but they can rarely produce any arguments or evidence. Yet Huxley provides an elegant summary of Darwin's ideas which makes an excellent basis to all teaching of evolution[20]:

1. the tendency for the numbers of all organisms to increase geometrically;
2. yet numbers are usually constant;
   (a) therefore there is a struggle for existence;
3. variation between individuals, some being better adapted than others;
   (b) therefore there will be natural selection;
4. some of this variation is inherited;
   (c) therefore the effects of differential survival accumulate, i.e. there will be permanent change in the populations.

Around this can be woven all other considerations and evidence[12, 14, 27, 28]. Curiously enough apart from studies on chromosome behaviour particularly the selective elimination of unbalanced genotypes[9], there has been little direct evidence on natural selection until recently. But now the accounts of industrial melanism[22], shell colour in snails[26] and the development of resistance to insecticides in insects[25] and to antibiotics in micro-organisms provide excellent and topical examples. The random nature of mutation and the lack of any connection between its origin and the demands of natural selection is best shown in micro-organisms. A good example is the development of resistance to bacteriophage in *Escherichia coli*[13]. This can lead on to a general consideration of Lamarckism. Here although there are many famous experiments that can be discussed, the best arguments are perhaps the evidence of the genes themselves, their inability to be altered directly and adaptively, as shown by countless genetical experiments that will have been discussed earlier. In such a discussion the significance of the findings of genetics to Darwin's theory can be explained; this will allow a short but pertinent digression on the history of Darwin's idea[6,14].

All this, however, provides a rather academic view of evolution. The exciting thing is that schoolchildren may observe it happening under their noses[5].

It is a process that can occur in 5 years and not 5 million, and populations of animals and plants near every school in town or country can be used to show adaptive evolutionary changes. Some possible examples which can be used are melanism and other characters in moths[14, 22], shell colour in snails[26, 27], climatic races in plants[7], heavy metal poison tolerant races in grasses[21]. In every species examples are to be found. Even the ubiquitous weed, annual meadow grass, *Poa annua*, provides excellent material. Animal material can be caught and examined in the field; plant material (seeds, tillers, etc.) can be brought back and grown in the school garden.

All that remains is a brief look at man. A photograph of human chromosomes[10] will remind a class very forcibly of man's affinities with other organisms and is a good starting point. The class can be tested for colour blindness and it is all the more interesting if twins can be included. But variation for most characters in human beings is continuous and its inheritance is therefore not so simple. But twins provide remarkable evidence of the importance of genetic influence on human make up and a pair of co-operative identical twins at school would certainly enliven a biology lesson. The end point of consideration of both man in particular and genetics in general can be a discussion of evolution in man at the present day[8, 15]. Eugenics, National Health Services, what can be achieved by artificial selection in other species, e.g. the breeds of dog, and the effects of atomic radiation[1, 2] and smoking can all contribute to an entertaining discussion giving point to the importance and wide implications of the subject.

REFERENCES

1. ALEXANDER, P. 1957. *Atomic Radiation and Life.* Penguin Books, London.
2. AUERBACH, C. 1956. *Genetics in the Atomic Age.* Oliver & Boyd, Edinburgh.
3. BEGG, C. M. M. 1959. *An Introduction to Genetics.* English Universities Press, London.
4. BRADSHAW, A. D. 1962. Brief Notes (see below).
5. BRIERLEY, J. K. 1961. Some suggestions for the teaching of evolution

in the field, garden and laboratory. *The School Science Review.* 148, 401.

6. CARTER, G. S. 1957. *A Hundred Years of Evolution.* Sidgwick & Jackson, London.
7. CLAUSEN, J. 1951. *Stages in the Evolution of Plant Species.* Cornell University Press, New York.
8. DARLINGTON, C. D. 1953. *The Facts of Life.* Allen & Unwin, London.
9. DARLINGTON, C. D. 1958. *Evolution of Genetic Systems.* Oliver & Boyd, Edinburgh.
10. DARLINGTON, C. D. AND LA COUR, L. F. 1962. *The Handling of Chromosomes* (4th Ed.). Allen & Unwin, London.
11. DARLINGTON, C. D. AND MATHER, K. 1949. *The Elements of Genetics.* Allen & Unwin, London.
12. DOBZHANSKY, T. 1951. *Genetics and the Origin of Species.* Columbia Univ. Press, New York.
13. DOBZHANSKY, T. 1955. *Evolution, Genetics and Man.* Wiley, New York.
14. DOWDESWELL, W. H. 1958. *The Mechanism of Evolution.* Heinemann, London.
15. DUNN, L. C. AND DOBZHANSKY, T. 1952. *Heredity, Race and Society.* Mentor Books, New York.
16. FORD, E. B. 1938. *The Study of Heredity.* Butterworth, London.
17. GOLDSCHMIDT, R. B. 1952. *Understanding Heredity.* Chapman & Hall, London.
18. GOLDSTEIN, P. 1957. *Genetics Made Easy.* Rider, London.
19. HASKELL, G. 1961. *Practical Heredity with Drosophila.* Oliver & Boyd, Edinburgh.
20. HUXLEY, J. S. 1942. *Evolution, the Modern Synthesis.* Allen & Unwin, London.
21. JOWETT, D. 1958. Populations of *Agrostis* spp. tolerant of heavy metals. *Nature, London,* **182,** 816.
22. KETTLEWELL, H. B. D. 1956. Further selection experiments on industrial melanism in the Lepidoptera. *Heredity,* **10,** 287.
23. MCLEISH, J. AND SNOAD, B. 1958. *Looking at Chromosomes.* Macmillan, London.
24. MATHER, K. 1957. *Genetics for Schools.* John Murray, London.
25. NEWMAN, J. F. 1957. Resistance to insecticides. *Outlook on Agriculture,* **1,** 235.
26. SHEPPARD, P. M. 1951. Fluctuations in the selective value of certain phenotypes of the polymorphic land snail. *Cepea nemoralis* (L.), *Heredity,* **5,** 125.
27. SHEPPARD, P. M. 1958. *Natural Selection and Heredity.* Hutchinson, London.
28. SMITH, J. M. 1958. *The Theory of Evolution.* Penguin Books, London.

MOST RECENT BOOKS

BONNER, D. M. 1961. *Heredity*. Prentice-Hall, New Jersey.

CARTER, C. O. 1962. *Human Heredity*. Penguin Books, London.

WALLACE, B. AND SRB, A. M. 1961. *Adaptation*. Prentice-Hall, New Jersey.

II BRIEF NOTES

SEEDLING CHARACTERS

LESLIE K. CROWE
Botany School, Oxford

Character differences visible in young seedlings are a space and time saving means of demonstrating Mendelian principles and sampling methods to beginners. They can also be used to show linkage, the special ratios arising in tetraploids, the correlation of variation in different structures, cytoplasmic inheritance, the effects of environment on viability and fertility, and the application of genetics to practical horticulture. The table on pages 92–93 shows some easily available examples. Where no commercial varieties are listed, seeds may be obtained from the Botany School, Oxford.

The general method used is to self-fertilise the heterozygous plants—produced by crossing or mutation—and examine the segregation in the seedlings which have the character of an F_2. Special examples are:

(i) The tetraploid *Rubus* in which *GGgg* gives *gggg* pure recessive (21 : 1).

(ii) The golden *Pelargonium* which is heterozygous and gives both homozygotes, green and white. This mutation exists not only in the non-chimaera type but also in a chimaera variety (Crystal Palace Gem) where only the two top layers have mutated to give the heterozygote; the bottom non-breeding layer is homozygous green.

Further, we should add two points of interest. The *Matthiola* heterozygote is one used in horticultural practice as a means of marking seedlings which are going to give double-flowered plants. And the rogue tomato is a cytoplasmic mutation arising under special temperature conditions which it is the purpose of tomato growers to avoid. The plants are infertile but the condition is not in the ordinary sense hereditary.

REFERENCES

1. C. D. CLAYBERG *et al.* 1960. *J. Hered.*, **51,** 167–174.
2. G. HASKELL. 1961. *Bot. Rev.*, **27,** 282–421 (Pleiocotyly).
3. D. LEWIS. 1953. *Heredity*, **7,** 337–359 (Rogues).
4. M. B. CRANE AND W. J. C. LAWRENCE. 1954. *The Genetics of Garden Plants* (4th Ed.). Macmillan, London.
5. A. J. BATEMAN. 1956. *Heredity*, **10,** 257–261.
6. C. D. DARLINGTON AND M. B. CRANE. 1932. *Nature*, **129,** 869.
7. R. A. TILNEY-BASSETT. 1963. *Heredity*, **18** (in the press).
8. K. F. THOMPSON. 1962. *Heredity*, **17,** 598.
9. E. H. COE. 1959. *Am. Nat.*, **93,** 381–382.

TABLE I

Seedling Characters

(dominant alleles marked *)

Species and References	*Character pair*	*Stage*	*Inheritance*	*Trade Varieties*
Tomato (1) (*Lycopersicum esculentum*)	red*—green	hypocotyl	1 gene	not in trade
	hairy*—hairless	hypocotyl	1 gene	not in trade
	normal*—potato	1st leaf	1 gene	not in trade
	normal—lanceolate	cotyledons	1 gene: het. intermediate, hom. lethal after germination	
(2)	1, 2 or 3	cotyledons	polygenic	Clucas 99
(3)	origin of the rogue mutation	2nd leaf	cytoplasmic; max. nos. at 30°C. and with 7 hrs. low intensity light	Ailsa Craig
Radish (4) (*Raphanus sativus*)	red—white	hypocotyl	1 gene: het. purple	Scarlet Globe,
	round—long		1 gene: het. intermediate	Icicle
Wallflower (5) (*Cheiranthus cheiri*)	red*—white	hypocotyl	1 gene: correlation with red and yellow flowers	Vulcan, Cloth of Gold
Blackberry (6) (*Rubus*)	glandular*—eglandular	cotyledons	1 gene: tetraploid: GGgg approx. 21:1, correlation with thorny and thornless stems	John Innes
Brompton Stock (4) (*Matthiola incana*)	dark*—light green	cotyledons	1 gene. Linkage with (i) pollen lethal gives 1:1; (ii) alleles for single and double flowers. Max. expression 9°C.	Hansen's vars.
Snapdragon (*Antirrhinum majus*)	straight*—twisted	Stems approx. 2″	1 gene	not in trade
	green*—yellow	1st leaf	1 gene	not in trade
	round*—narrow	1st leaf	1 gene	not in trade

TABLE I—*continued*

Species and References	*Character pair*	*Stage*	*Inheritance*	*Trade Varieties*
Nasturtium (4) (*Tropaeolum majus*)	green*—variegated	1st leaf	1 gene	not in trade
Wintercress (*Barbarea vulgaris*)	green*—variegated	1st leaves	1 gene	not in trade
Arabidopsis thaliana	green*—variegated	cotyledons	1 gene	not in trade
Pelargonium zonale (7)	green*—white	young seedlings	1 green: 2 het. gold: 1 white (sub-lethal)	Golden Crampel
Marrow stem Kale (8) (*Brassica oleracea*)	red—green	hypocotyl	1 gene. Het. purple Linked with incompatibility gene. Max. expression 19°C.	not in trade
Maize (9) (*Zea mays*)	dull*—glossy	young seedlings	1 gene	not in trade
	diploid—haploid	young seedlings	Probably 2 genes: Stock 6 gives 3% haploids when selfed. When crossed on to glossy strain (gg × GG) gives 2·3% glossy seedlings which are haploids (g)	not in trade

LEAF MARKINGS IN *TRIFOLIUM REPENS*

W. Ellis Davies
Welsh Plant Breeding Station
Aberystwyth

The leaf-marking series in white clover provide useful material for the study and demonstration of multiple allelic series. Such material is not common in the higher plants, and is especially useful as it can be demonstrated at any time of the year. In addition, white clover is one of our most common plants growing in a wide variety of soils and situations, occurring as wild populations in uncultivated pastures, and used as bred varieties in sown pastures.

The leaf marks are of two kinds, the white and the reddish patterns.

White patterns

These are designated as the V series, after the most frequent form of the pattern. The paler areas of the patterns are due to the palisade cells being smaller, shorter and more irregularly shaped, and with more inter-cellular spaces than the adjacent cells (Carnahan *et al.* 1955). Seven of the pattern alleles in the homozygous condition are shown in column one of Plate 3. Absence of marking (vv) is recessive to presence of marking (e.g. V^lV^l), so that the F_1 of a cross $V^lV^l \times vv$ has a leaf-mark and has the genetic constitution V^lv. The F_2 progenies segregate into 3 plants with, to one without, leaf marks, while backcrosses of the F_1 to the recessive parent ($V^lv \times vv$) segregate in a 1 : 1 ratio (Brewbaker 1955).

Crosses of different marker alleles result in a compound mark in which the effect of both alleles can be observed when they are on a different position on the leaf. When both alleles produce an effect in the same position, or close together, on the leaf one mark may mask the expression of the other. In such cases the shape and size of the leaflet become important; a large, long leaflet facilitates identification, while a small rounded leaflet hinders it.

Thus the marking alleles show no dominance although in the presence of the allele V^{by} the expression of V^{f} and V^{ba} is often reduced. In the present work the leaf marking locus has behaved as a single unit, but by using gamma irradiation, Davies and Wall (1960) have been able to break up the V^{by} locus into a yellow point, a white broken V, and other forms. This suggests that this locus is complex, although no cross-over types have been observed in actual crosses. Certain anthocyanin patterns which are observed to be associated with the V^{by} allele have not beens een in association with the other V alleles (Plate 4, 7–10).

The V patterns illustrated are distinctive types which give reasonably clear visual separation, but in practice considerable plant-to-plant variation occurs in position, size and intensity including a range between the extreme of low V^{l} to high V^{h}. An intermediate allele V^{i} has been listed by Brewbaker and Carnahan (1956).

Reddish patterns

These patterns are due to anthocyanin. They include the R allelic series (Plate 4, Nos. 1–6) which are independent of the V locus, and a second group which have been found only in association with the V^{by} locus (Plate 4, Nos. 7–10).

The presence of anthocyanin (e.g. $R^{m}R^{m}$) is dominant to its absence (rr). One anthocyanin allele may mask the expression of another so that it may be difficult or impossible to detect the presence of two alleles by visual observation. The presence of an R^{f} allele in $R^{l}R^{f}$ or $R^{m}R^{f}$ genotypes has to be confirmed by test crosses.

The inheritance of the second group of anthocyanin patterns is not clearly understood and they are only included to distinguish them from the R series. So far they have been found only in V^{by} genotypes.

In the R series pigmentation occurs in the upper and/or lower epidermis, and is intensified at temperatures below 10°C.

Most patterns are however visible throughout the year (Carnahan *et al.* 1955), and therefore differ from the general reddening that often occurs at low temperatures.

Visibility falls into three grades:

Patterns 1–7 are visible throughout the year.
Patterns 8 and 9 are often faint and difficult to detect.
Pattern 10 can be observed only in winter and spring and disappears in summer.

Uses for teaching

The white pattern V-series can be used in teaching to demonstrate

(i) Multiple allelic series, only two alleles occur in one plant. This can be set up as a permanent live demonstration suitable for a botanical garden or grown in pots.

(ii) Mendelian genetics, to illustrate dominance, mosaic

PLATE 3 (opposite)

THE V-LEAF MARKINGS OF WHITE CLOVER

First column homozygotes

v	Absence of leaf mark—recessive to all others
V^l	Low V-mark in lower half of leaflet, size and position variable
V^h	High V-mark extending into upper half of leaflet
V^f	Filled in area within V is marked
V^{ba}	Basal, narrower, longer and fainter than V^f
V^b	Broken, point of V is absent
V^{by}	Broken yellow, point of V is yellow, arms are white

The remainder are compound leaf marks, each of two alleles arranged in a diallel series. There is no dominance, both marks in a compound mark can be observed if they are in a different position on the leaf. When they are in the same position, or close together, the expression of one allele is often masked by the other. Thus it is impossible to identify V^l in a V^lV^f compound, and only in favourable circumstances can the second allele be observed in V^fV^{ba}, $V^{by}V^l$, $V^{by}V^h$ compound marks.

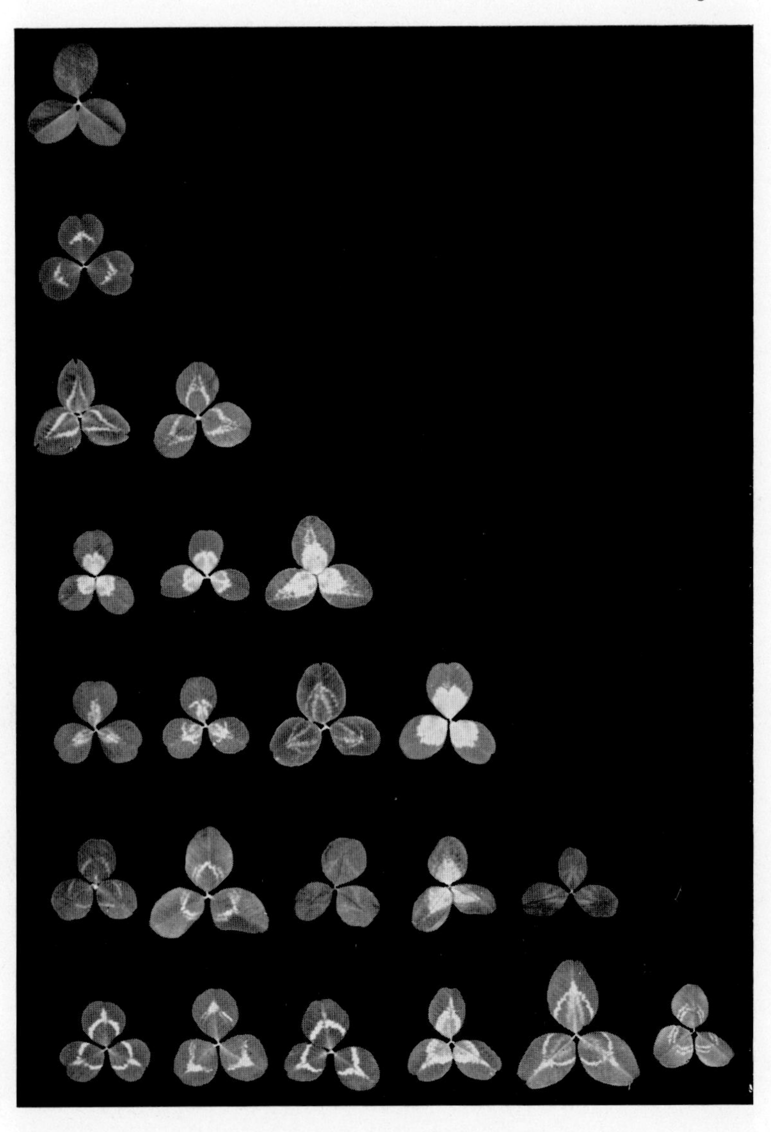

V

V^{l}

V^{h} $V^{l}V^{h}$

V^{f} $V^{l}V^{f}$ $V^{h}V^{f}$

V^{ba} $V^{l}V^{ba}$ $V^{h}V^{ba}$ $V^{f}V^{ba}$

V^{b} $V^{l}V^{b}$ $V^{h}V^{b}$ $V^{f}V^{b}$ $V^{ba}V^{b}$

V^{by} $V^{l}V^{by}$ $V^{h}V^{by}$ $V^{f}V^{by}$ $V^{ba}V^{by}$ $V^{b}V^{by}$

PLATE 4

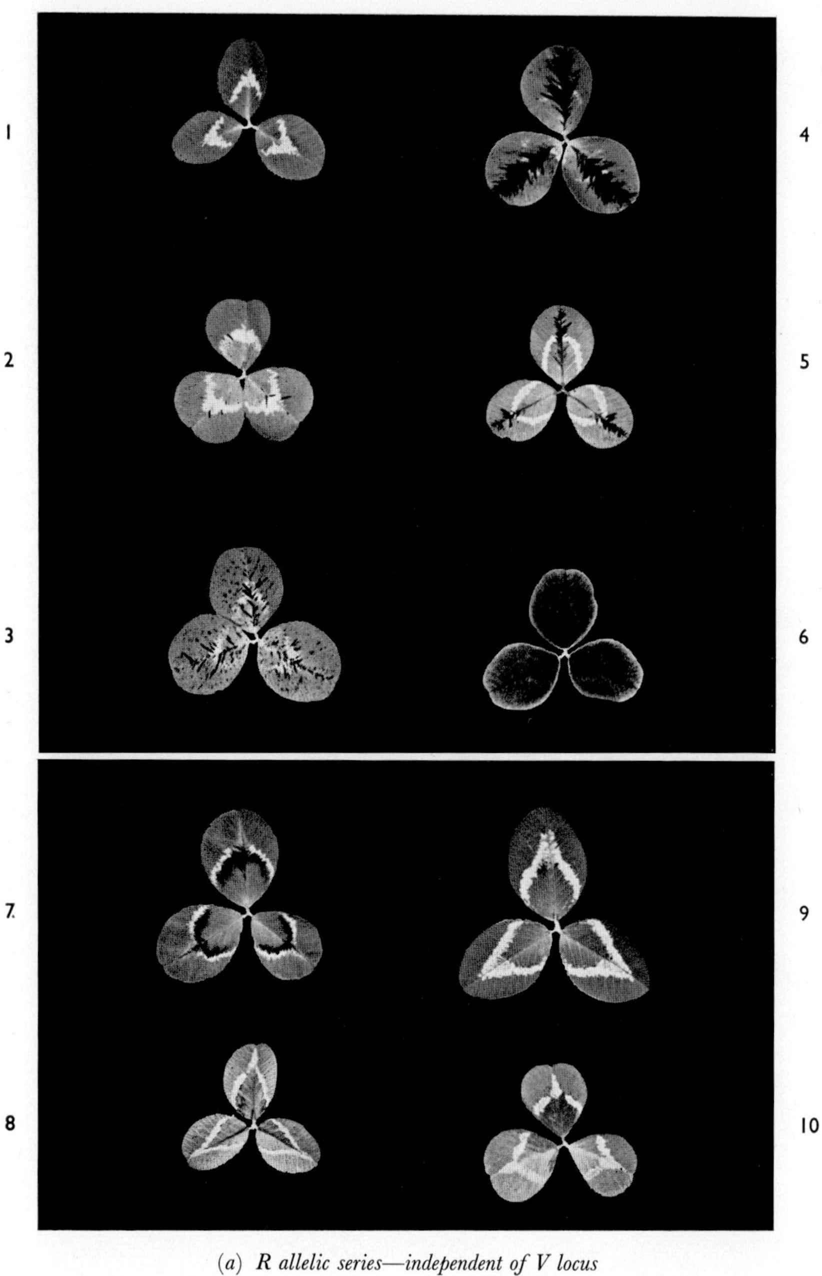

(*a*) *R allelic series—independent of V locus*

	R allele	*V allele*		*R allele*	*V allele*
1	r	V^{l}	4	R^{m}	V^{l}
2	R^{f}	V^{l}	5	R^{m}	V^{by}
3	R^{fa}	V^{l}	6	R^{l}	$V^{l}V^{l}$

(*b*) *Anthocyanin patterns associated with V^{by} locus (inheritance not fully understood)* dominance (compound markings), segregation ratios, 1 : 1, 1 : 2 : 1, 1 : 1 : 1 : 1, and 3 : 1.

(iii) Population genetics, the frequency of a distinctive allele, e.g. v, in wild populations from uncultivated pastures, sports grounds, and waste places can be observed and compared with those from cultivated pastures.

The leaf marks described occur in wild and cultivated forms in varying frequencies. There is a tendency for the mark to be more pronounced in populations nearer the centre of origin which are also larger-leaved than indigenous white clover in Britain.

PLATE 4 (opposite)

RED (ANTHOCYANIN) PATTERNS IN WHITE CLOVER

1. r Absence of any anthocyanin, recessive to presence.
2. R^f Flecks—a few small red flecks sometimes only on a few leaves on the plant. A frequent allele.
3. R^{fa} Frequent flecks on all leaves upper and lower surface. An infrequent allele.
4. R^m Midrib—anthocyanin restricted to area along midrib. Infrequent allele.
5. R^m When in association with V^{by} the intensity of anthocyanin is reduced in the area inside the V mark.
6. R^l Red leaf, all except the margins of the leaves are red. Often found in Botanical Gardens and described as var. *purpureum*, but rare in wild populations.
7. Red V. This occurs as a well-marked V diffusing into the rest of the leaflet, especially under low temperatures. Hovin & Gibson (1961) who supplied this material propose the symbol V_2^r for this gene.
8. Thin fringe of red on the inside of V^{by}, but in $V^{by}V^l$, or $V^{by}V^h$ compounds, the whole of the area occupied by V^l or V^h may be pigmented.
9. Red spot at point of V, infrequent.
10. Seen only at low temperatures; inside V on central leaflet, but sometimes in adjoining margins of side leaflets. Frequent in cultivars.

Propagation

White clover is a perennial and spreads by surface creeping stolons which root readily at the nodes. Vegetative propagation is quite easy, and cuttings can be taken at any time of the year. Plants in pots should be re-potted every year.

White clover is practically self-sterile, and in nature cross-pollination is effected by bees. The plants flower out of doors from May to July, and artificial hybridisation is best done in a bee-proof glasshouse. It is not necessary to emasculate, unless very exact results are required. A suitable technique for this has been described by Williams (1954).

Collections of material can be built up from commercial varieties. Plant material is maintained at the Welsh Plant Breeding Station and is generally available in small quantities. A cultivar, S.100 Nomark (vv genotype), is being marketed, and marker-lines for use in experimental work of R^m and R^l genotypes are being developed.

REFERENCES

BREWBAKER, J. L. 1955. V-leaf markings of white clover. *J. Hered.* **46,** 115–123.

BREWBAKER, J. L. AND CARNAHAN, H. L. 1956. Leaf marking alleles in white clover. Uniform Nomenclature. *J. Hered.* **47,** 103–104.

CARNAHAN, H. L., HILL, H. D., HANSON, A. A. AND BROWN, K. G. 1955. Inheritance and frequencies of leaf markings in white clover. *J. Hered.* **46,** 109–114.

DAVIES, D. R. AND WALL, E. T. 1950. Induced mutations at the V^{by} locus of *Trifolium repens. Heredity.* **15,** 1–15.

HOVIN, A. W. AND GIBSON, P. B. 1961. A red leaf marking in white clover, *J. Hered.* **52,** 295–196.

WILLIAMS, W. 1954. An emasculation technique for certain species of *Trifolium. Agronomy J.* **46,** 182–184.

CYANOGENESIS IN *TRIFOLIUM REPENS*

J. G. Pusey
Botany School, Oxford

Introduction

White clover, *Trifolium repens* L., is grown as a forage crop throughout the temperate zones. Spontaneous populations are now found throughout this area, as well as in its original home, Europe and Western Asia. It is apparently an allotetraploid, and is an outbreeder, with a system of incompatibility alleles similar to that found in *Nicotiana*.

Both in wild and cultivated populations there is a genetic polymorphism controlling cyanogenesis. The moment that they are removed from the plant, the leaves of some plants start to release easily detectable amounts of hydrogen cyanide. This is the result of the action of an enzyme on two glucosides. The presence of the enzyme and the presence of the glucosides are both under simple genetic control.

Hereditary Control of Cyanogenesis

By the simple test for the production of HCN which is described below, populations of *T. repens* can be divided into three classes:

(i) Cyanogenic, and giving a rapid yield of HCN
(ii) Cyanogenic only after more prolonged incubation
(iii) Entirely acyanogenic

Cyanogenesis depends primarily on the presence of a mixture of two 'cyanogenic glucosides', linamarin and lotaustralin. This is controlled by a dominant allele *Ac* (*ac*, acyanogenic). Class (iii) is therefore *ac ac*.

Class (ii) lack the enzyme linamarase, whose presence in

Class (i) is determined by a dominant allele *Li*: Class (ii) is therefore *Ac_ li li*, while Class (i) is *Ac_ Li_*.

Class (iii), which contains the *ac ac* plants, can be further divided on the basis of the plants' enzyme phenotypes. If a glucoside source is added to the leaves being tested, only *ac ac Li_* plants will catalyse its rapid breakdown.

These two loci are unlinked, and no linkage has been demonstrated for either with other known loci.

The plants can thus be referred to four phenotypes, which will segregate in the proportions shown, in an F_2 derived from crosses among an F_1 generation of *Ac ac Li li* plants.

9 *Ac_ Li_*	3 *ac ac Li_*
Cyanogenesis in detached leaves is spontaneous and rapid.	Acyanogenic: but will catalyse rapid breakdown of added glucosides.
3 *Ac_ li li*	1 *ac ac li li*
Cyanogenesis slow, unless enzyme added.	Acyanogenic, and lacking enzymic activity.

The Cyanogenic Glucosides

$$\begin{array}{l} CH_3 \quad C_2H_5 \\ \quad \diagdown \; \diagup \\ \quad\; C - O - \text{Glucose} \\ \quad\; | \\ \quad\; CN \end{array}$$

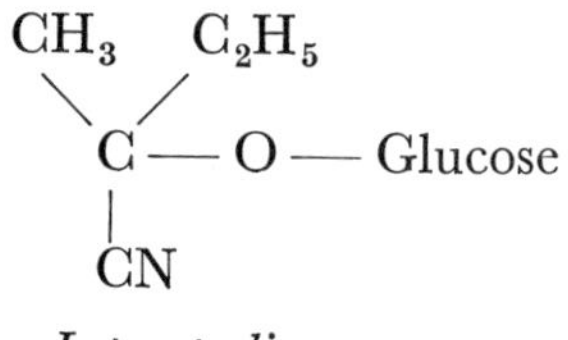

Lotaustralin

The compound illustrated, *lotaustralin*, represents about three-quarters of the cyanogenic glucosides in *T. repens*. Hydrolysis removes the glucose, and the aglycone breaks down to give HCN and methyl-ethyl ketone. In the other glucoside, linamarin, a methyl group replaces the ethyl group.

Glucoside Test

A simple test for cyanogenesis can be carried out to distinguish the *Ac_ Li_* phenotype from the others, making use of the colour produced in picric acid derivatives by HCN.

A nearly saturated (0·05 M.) solution of picric acid is neutralised by the addition of $NaHCO_3$, and filtered after standing. 'Sodium picrate papers' are prepared by soaking filter papers in this solution. These papers can be dried and cut into strips, and kept indefinitely in a closed container.

In a 2 in. by $\frac{5}{8}$ in. glass tube, place two leaves (newly expanded, if possible) of the plant to be tested. Add a drop of water, and macerate with a glass rod. Add two drops of toluene, and cork the tube firmly, placing a short strip of sodium picrate paper so that the cork holds it clear of the contents and the sides of the tube. Incubate at 40°C. for two hours. In a positive test there should be a marked change from a clear yellow to a reddish brown. Only *Ac_ Li_* plants will give a positive reaction.

Non-enzymic hydrolysis will also occur under these conditions, and takes ten or more times as long as the enzymic hydrolysis. If the papers are examined after not less than 24 hours, the procedure above can be used to test directly for the presence of glucosides. This will identify all the *Ac_* plants as positive.

The *Ac_ li li* plants may be identified by the change in reaction from negative to positive between the earlier and later observations. They may be identified more quickly by a test in which leaves of the complementary phenotype *ac ac Li_* are added as a source of enzyme, so that a quick positive reaction is given by the mixture. Maceration is clearly essential in this test.

Enzyme Test

A test for the presence of enzyme is needed to distinguish *ac ac Li_* and *ac ac li li* plants. This requires the addition of a source of glucosides. Pure glucosides can be prepared, but a crude extract of glucoside-containing leaves is normally adequate (or *Ac_ li li* leaves may be used). The glucosides are present only in the leaf lamina.

The extract should ideally be prepared from *Ac_ lili* plants, but in fact *Ac_ Li_* plants can be used if the leaves are quickly autoclaved to inactivate the enzyme. *Note:* 25 per cent. of the glucosides can be broken down within 15 minutes of the leaf being removed from the plant. This may be the 25 per cent. which is present in the cytoplasm, the remainder being in the vacuoles.

Autoclave at 110°C. (up to 10 lb. excess pressure) for 15 minutes. More intense autoclaving may reduce the yield. Extract 100 fair-sized leaves into about 25 ml. of water; the brown filtered extract can then be used as a glucoside source. It may keep its activity for several days, but this should not be relied on. Two drops of this preparation, in place of the drop of water, should give a reasonable yield of HCN when added to leaves of a plant containing the enzyme.

Value in Teaching

This is one of the few cases where one can demonstrate a 'one gene—one enzyme' relationship in higher plants. Linamarase activity is inherited in the simplest possible way, its presence depending on a dominant allele at a single locus. The presence or absence of the two glucosides, probably the result of the presence or absence of a specific enzyme, is also a good example of a gene whose expression is physiologically simple.

The whole system can be used as an example in which a dihybrid cross will show a 9 : 7 ratio of cyanogenic to acyanogenic plants segregating in the F_2. By different tests, or combinations of tests, this F_2 could also be analysed as a 12 : 4, a 9 : 4 : 3, or a 9 : 3 : 3 : 1 segregation.

The first step in introducing the material might be to demonstrate by the simple test that some plants were not cyanogenic. Then, by mixing together leaves of *ac ac Li_* and *Ac_ li li* plants, one could show that these two different genotypes, both inactive when tested alone, complement one another *in vitro* to give a positive test for cyanogenesis. This physiological evidence shows that there is independent variation in the presence of two factors, both essential for the reaction. It may be presented alongside the genetic evidence, from the 9 : 7 segregation, that two loci are involved. These two pieces of evidence both indicate the presence of two component systems which interact to produce cyanogenesis. The rest of the evidence serves to identify the particular roles of the two loci, and it then becomes clear that the physiological and genetical approaches both analyse the system into the same two parts.

The students' grasp of the situation may be examined by asking them to arrange tests to determine the phenotypes of 'unknown' clones.

Ecological Significance

It seems likely that these phenotypes have different selective values in nature. Many small animals can discriminate, preferring to eat acyanogenic plants. The cyanide produced by

T. repens is apparently not concentrated enough to harm or kill herbivorous mammals, though it can affect their intestinal fauna.

Large population samples from different sites show differing frequencies of the four phenotypes. But, in any population, if one dominant allele is frequent, the other is also found to be frequent, and when one is rare, both are rare. Also, in populations where the frequencies of the dominant alleles are high, the actual quantity of glucosides produced by the *Ac*_ plants tends to be higher. So the gene frequencies at both loci, as well as a system of modifying genes, appear to respond together to a single selective pressure. This suggests that cyanogenesis is indeed the property of these alleles on which selection acts.

The distribution of gene frequencies follows a very regular pattern on the map. The recessive alleles are more frequent towards higher latitudes or altitudes—that is, towards colder regions. The distribution fits that of winter temperature. It appears that temperature extremes, especially extreme cold, are capable of activating the enzyme while the plant is still intact. The consequent release of HCN inhibits respiration and all energy-requiring processes dependent on it, and puts cyanogenic plants exposed to such cold conditions at an evident disadvantage, presumably outweighing the advantage they would otherwise obtain by deterring their predators. The species has become naturalised in North America, and there too a ratio cline is found, in the same direction as in the Old World, but less steep.

Large population samples, almost without exception, show that both alleles are present at each locus, and that none of the alleles are fixed. The incidence of extreme cold (harmful to cyanogenic plants) will be intermittent, and the incidence of selective predation (harmful to acyanogenic plants) may also be irregular, both in time and space. These two factors, and especially any irregularities in their incidence, may produce disruptive selection of such a kind as to make the polymorphism stable. It is also possible that the polymorphism is 'internally balanced' by some advantage of the heterozygotes.

This example is one of a double balanced polymorphism, with clinal variation in the local equilibrium reached by the gene frequency. A similar situation exists in *Lotus corniculatus*, which also shows genetic variation affecting cyanogenesis.

REFERENCES

ATWOOD, S. S. AND SULLIVAN, J. T. 1943 Inheritance of Cyanogenetic Glucoside and its Hydrolysing Enzyme in *Trifolium repens*. *J. Hered.*, **34**, 311–320.

BUTLER, G. W. AND BUTLER, B. G. 1960. Biosynthesis of linamarin and lotaustralin in White Clover. *Nature*, **187**, 780.

CORKILL, L. 1952. Cyanogenesis in White Clover (*Trifolium repens* L.). VI. Experiments with high glucoside, and glucoside-free strains. *N.Z. J. Sci. Tech.*, **34**, **A**, 1–16.

DADAY. H. 1954. Gene frequencies in wild populations of *Trifolium repens*. I. Distribution by Latitude. *Heredity*, **8**, 61.

DADAY, H. 1955. Cyanogenesis in strains of White Clover. *J. Brit. Grassl. Soc.*, **10**(3), 266–274.

DADAY, H. 1958. Gene frequencies in wild populations of *Trifolium repens*. III. World Distribution. *Heredity*, **12**, 169.

DADAY, H. 1962. White clover. Mechanism of natural selection. *Ann. Rept., Div. of Plant Industry, C.S.I.R.O., Canberra*, 1961–1962, p. xiii, p. 14.

HAYES, H. K., IMMER, F. R. AND SMITH, D. C. 1955. Methods of Plant Breeding (2nd Ed.). McGraw-Hill.

JONES, D. A. 1962. Selective eating of the Acyanogenic Form of the Plant *Lotus corniculatus* L. by various animals. *Nature*, **193**, 1109.

TETLEY, J. H. 1953. Inhibitions of Populations of *Haemonchus contortus* in Sheep fed on White Clover (*Trifolium repens*) high in Lotaustralin. *Nature*, **171**, 311.

THREE TEACHING PROJECTS

A. D. Bradshaw
Department of Agricultural Botany,
University College of North Wales, Bangor

Antirrhinum

Cultivated varieties of *Antirrhinum* provide excellent material for genetical and biochemical teaching. The following suggestions stem originally from the account by Dayton (1956) which provides further details.

Antirrhinum is particularly easy to grow and to cross. Every gardener is familiar with it and its propagation involves no problems. Crosses can be made very easily, by untrained people. Anthers are large and are easily removed from the flowers just before they open. Simple bagging precludes insects and therefore accidental cross pollination. A single successful pollination provides a large quantity of seed. Nearly all varieties are pure lines.

The varieties differ markedly in flower colours. These can be analysed by simple paper chromatography. The upper lips of two corollas should be ground up in a test tube in 1 ml. of 1 per cent. HCl in ethyl alcohol. About 4 drops of this extract should be placed on one spot on a piece of chromatographic paper, time being allowed for each drop to dry before the next is applied. The chromatogram should then be carried out using butanol: acetic acid: water, 6 : 1 : 2, as a solvent. While the elaborate descending methods can be used, simple ascending methods are excellent. A large sheet of paper can be spotted with many different extracts, the edges clipped together to form a cylinder, and then stood in the solvent in a suitable airtight container (accumulator jar, saucepan, etc.). Otherwise narrow strips with single spots can be suspended individually in corked boiling tubes. From such chromatograms the three main pigments affecting flower colour, cyanidin, pelargonidin and areusidin can be distinguished (details in Dayton, 1956).

The flower colours are made up from various combinations of these pigments, whose occurrence and distribution is controlled by major genes and whose intensity may be controlled by polygenic modifier complexes. Crosses therefore can be made to give simple and complex ratios (one factor, two factors, two factors with interaction, etc.). The differences in intensity can be utilised to demonstrate the inheritance of continuous varying characters. Table I gives the phenotypes and genotypes of a number of useful varieties, and Table II some possible

TABLE I

Genotypes and phenotypes of Antirrhinum *varieties* (Carter's seed)

	Genotype	*Areusidin*	*Cyanidin*	*Pelargonidin*
White	r B I	—	—	—
Yellow	r B I_A	α	—	—
Nelrose (glowing pink)	R b I	—	—	β
Royal Cerise (rich carmine)	R b I	—	—	α
Scarlet Flame (fiery scarlet)	R b I_A	γ	—	α
Guinea Gold (orange terracotta)	R b I_A	α	—	β
Mauve Beauty	R B I	—	β	—
Crimson	R B I_A	β	α	—

TABLE II

Possible crosses

Single factor	pink × mauve, pink × scarlet
Two factor	pink × crimson, scarlet × mauve
Modifier complex	pink × cerise, scarlet × terracotta
Two factor with epistasis	white × pink, yellow × crimson

Delila segregation can also be incorporated in some crosses

crosses. Apart from these there are a number of varieties with a colourless corolla tube (*delila* types) due to the single gene t.

The only drawback to the material is that it cannot easily be made to flower in the winter. However, the flowers, if dried rapidly, retain their colours excellently and can be permanently mounted on herbarium sheets. At the same time further flowers can be dried and stored for chromatographic analysis.

Heavy Metal Tolerance

The populations of some plant species found growing on the old workings of certain metalliferous mines have been shown to possess remarkable tolerances to heavy metal toxicity, quite different from normal populations of the same species. Such species include *Agrostis tenuis*, *Festuca ovina*, *Melandrium dioicum* and *Thlaspi alpestre*.

Populations of *Agrostis tenuis* found on different mines have been shown to possess different and specific tolerances to individual heavy metals—lead, copper and nickel (Jowett, 1958). Such populations provide excellent material in which evolutionary adaptation and differentiation in physiological characters can be demonstrated, as well as the selective processes involved.

The primary and major effect of the heavy metals concerned is to inhibit root growth, and the length of the roots provides the best character by which the characteristics of the populations may be determined.

Existing populations taken from mines as well as normal habitats should be tested as follows:

(i) Tillers should be collected at random from each population to be tested and grown on in standard potting soil for a period of about six weeks.

(ii) Tillers should be taken from the resulting plants. From each plant (or clone) one or two tillers should be allowed to root in distilled water containing $\frac{1}{2}$ gm./litre hydrated calcium nitrate, and an equivalent number of tillers from the same plant allowed to root in a testing solution similar to the first but containing one required metal in the following concentration: lead, as lead nitrate, 125 micro moles per litre; or nickel, as nickel sulphate, 10 micro moles per litre; or copper, as copper sulphate, 5 micro moles per litre. The tillers should be stood or suspended in beakers of the solution, until the roots are 5–10 cms. long in the controls. For critical work the solution should be changed regularly.

(iii) Indices of tolerance for clones or populations can be determined from the growth in the testing solution expressed as a percentage of the growth in the controls. Tolerant populations will have values between twice and ten times those of the intolerant.

The characteristics of seedling populations can be determined in an equivalent manner. Samples of seeds of populations to be investigated should be scattered on polythene gauze (obtainable from Henry Simon, Cheadle Heath, Stockport) floating on the solution in covered beakers. An alternative method is to stretch nylon net (e.g. old stocking material) over small beakers filled to the brim with solution. These beakers should be placed in large covered containers to prevent evaporation, but even so the level of the solution must be watched. If the experiment is carried out in a greenhouse the roots can be measured after three weeks.

The results from the seedling material can be used in various ways. They demonstrate the nature of the selective process (the lack of rooting and the death of the seedlings). If carried out critically they can be used to determine gene flow, and the genetic basis of the character (by analysis of pair crosses, etc.).

The Effects of Metal Ions on Cell Division

It has been briefly recorded (Levan, 1945) that many metal ions have marked effects on cell division, usually causing spindle inhibition as in c. mitosis. Such effects are of interest in several different directions.

Firstly, it appears that some ions and not others cause c. mitosis. This may be related to the particular chemical effect of the ion, especially with regard to protein metabolism.

Secondly, it is known (Jowett, 1958) that the effects of particular ions differ in different plants and in different populations of a single species. In *Agrostis* sp. some populations are for instance tolerant of remarkably high concentration of lead.

Thirdly, some metal ions notably aluminium and manganese are well known for their important effects on plant growth in acid soils (Hewitt, 1946). It has been shown (Hewitt, 1949) that different species have very different tolerances to them. Their cytological effects are not well known but Hewitt records that while aluminium causes a considerable reduction in root growth, manganese does not.

The following therefore make very satisfactory investigations in experimental cytology and ecology. Effects on overall growth, general antimitotic activity and specific effects such as c. mitosis should be recorded.

(i) The effect of aluminium, manganese, and a distinctly toxic metal such as lead, on root growth and mitosis of a single species.

(ii) The root growth and mitosis of contrasting species in relation to one of these metals, perhaps aluminium.

Suggested investigations:

Species	Metal	Concentration
Shallot, Oats, Barley	aluminium (as aluminium sulphate)	100 micro moles/litre
		200 ,, ,, ,,
Shallot	manganese (as manganese sulphate)	100 micro moles/litre
		500 ,, ,, ,,
		2500 ,, ,, ,,
Shallot	lead (as lead nitrate)	25 micro moles/litre
		125 ,, ,, ,,
		626 ,, ,, ,,

REFERENCES

DAYTON, T. O. 1956. The inheritance of flower colour pigments. I. The genus *Antirrhinum*. *J. Genet.*, **54**, 249–260.

HEWITT, E. J. 1946. The resolution of the factors in soil acidity. *Ann. Rep. Long Ashton Res. Sta.* 1945, 51.

HEWITT, E. J. 1949. The resolution of factors in soil acidity. IV. The relative effects of aluminium and manganese toxicities on some farm and market garden crops. *Ann. Rep. Long Ashton Res. Sta.*, 1948, 58.

JOWETT, D. 1958. Populations of *Agrostis* spp. tolerant of heavy metals. *Nature, London.* **182,** 816.

LEVAN, A. 1945. Cytological reactions induced by inorganic salt solutions. *Nature, London.* **156,** 751.

A *DROSOPHILA* POPULATION CAGE FOR CLASS EXPERIMENTS

W. J. Whittington
University of Nottingham, School of Agriculture, Sutton Bonington

The cage consists of components of three kinds:

1. A 370 ml. plastic beaker with lid made by the Paper Cap Manufacturing Co., Feltham, Middlesex.
2. Three $2 \times \frac{1}{2}$ in. specimen tubes containing *Drosophila* medium (Darlington and La Cour, 1962).
3. A drilled muslin-covered cork for aeration.

Four holes are drilled in the lid to take the cork and the three specimen tubes. The holes for the tubes should be equally spaced around the hole for the cork which should be in the centre of the lid. The tubes are placed with their open ends inside the beaker. Their closed ends protrude from the lid for about 1 inch and the cage is made to stand on the tripod they form.

Flies may be introduced by removing, temporarily, either the lid or one of the tubes. Three to ten flies of each sex is a suitable number. A known frequency of wild-type and mutant genes can be given by using heterozygous flies or mixtures of homozygotes. Changes in gene frequency are followed by classifying the entire population (75–150 flies) at given time intervals, e.g. every two weeks. To accomplish this:

1. The cage is inverted, the lid loosened and the flies tapped into the beaker.
2. The lid is removed and the beaker re-inverted quickly over a funnel leading to an etherising vessel. The flies are tapped into the vessel, classified and returned to the cage.

The cages can be changed when dirty and the main points to watch are the following:

1. Medium is not too fluid.
2. Aeration is adequate.
3. Tubes are changed at regular intervals to reduce the risk of genetic drift.

The apparatus can be used to demonstrate selection at work within the population. The probable rate of disappearance of a mutant can be forecast approximately from the degree of difference between it and the wild-type. Thus, vestigial-winged flies (vg) are expected to be at a greater disadvantage with respect to wild-type than brown-eyed flies (bw).

Clearly, more complex experiments can be carried out where several different mutants are followed simultaneously or where suitably marked inversions or translocations are available (Whittington and Peat, 1960). The analysis of the data illustrates to students the principles of natural selection.

REFERENCES

DARLINGTON, C. D. AND LA COUR, L. F. 1962. The Handling of Chromosomes. Allen and Unwin 4th Ed.

WHITTINGTON, W. J. AND PEAT, W. E. 1960. The behaviour of the Xasta (Xa) mutant of *D. melanogaster* in populations. *Dros. Inf. Ser.*, **34**, 110.

A PRACTICAL EXAMINATION MODEL

H. L. K. Whitehouse
Botany School, Cambridge

In a practical examination, candidates are asked to prepare a model of one of the bivalents seen in a microscope preparation showing metaphase I of meiosis. The model is to be made by threading plastic tubing of two colours on to wire, twisting it to the appropriate shape, and labelling the centromeres. The question is found to reveal deficiencies in certain of the candidates' understanding of the structure of bivalents, such as would not have been apparent if they had merely been asked for an annotated drawing.

III. SOURCES OF MATERIALS

Introduction

The present compilation is arranged in two sections, the first being a list of material and the second a list of the contributors to the scheme. The list of material is arranged under the specific names of the organisms concerned, and includes a brief statement of their values in teaching. The last column of this list, headed "Source", refers to the contributor by his number in the final section.

1. LISTS OF MATERIAL

A. Bacteria

Organism	*Description and use*	*Source*
Escherichia coli	K12—fertility and mutant types—conjugation, transduction	2a, 6, 8b, 11, 18
Escherichia coli	B—host for "T" bacteriophage	2a, 6, 11
Salmonella typhimurium	LT2 various mutants—transduction, mutation (reversions), biosynthetic pathways	2a, 4, 6, 8b, 11
Klebsiella pneumoniae	capsulated and non-capsulated—differences in phage sensitivity	8b
Bacillus subtilis	various mutants—transformation	6

B. Bacteriophage

T4	fine structure	2a, 6
PLT–22	transduction	2a, 4, 6, 11
λ	transduction, phage crosses	6
Note on A and B:	the strains provided will be limited to those used for demonstrating classical experiments	

C. Fungi

Organism	Description and use	Source
Aspergillus nidulans	wild type and mutants—crosses, production of diploids and mitotic analysis	6, 14, 18
Coprinus lagopus	wild type and auxotrophs—segregation and linkage in haploid progenies, tetrapolar incompatibility, complementation in dicaryons	13
Neurospora crassa	STA (wild type) and 37–402a (pale ascospore mutant)—Mendelian segregation in asci	9a
Neurospora crassa	*albino a*—white conidia—can be scored in colonies on plates	9a
Neurospora crassa	various auxotrophs—especially arginine requirers—demonstration of growth response and arginine pathway	5
Neurospora crassa	*asco* (35A) × wild (Abb 12a)—distance between *asco* and centromere. Effect of temperature on recombination	5
Neurospora crassa	*asco* (CUa) × Abb/12/L–9.1A *asco* (CUa) × Abb/12/L–9.2A —segregation for differential recombination frequences	5
Neurospora crassa	strains showing heterocaryon compatibility within but not between wild strains. Also trans test for functional allelism	5
Schizophyllum commune	shows tetrapolar incompatibility in appropriate matings by presence or absence of clamp connections	5
Saccharomyces cerevisiae	wild types "a" and "α", mutants ad_1 and ad_2—complementation various auxotrophs for the demonstration of linkage	2a, 11 2a
Schizosaccharomyces pombe	homothallic and heterothallic mating types—mating and ascospore formation. Various auxotrophs for random ascospore analysis	8b
Sordaria fimicola	types with black and white ascospores—mapping colour locus and centromere	2a, 11
Ustilago maydis	wild type and auxotrophic haploids and heterozygous diploids—replica plating, segregation, heterothallism, pathogenicity	9a

D. Animals

Organism	*Description and use*	*Availability*	*Source*
Drosophila melanogaster	wild types and several mutants—Mendelian segregation, biometry, salivary glands	one culture of each	17a
Gallus domesticus (fowl)	Light Sussex showing "talpid" lethal effect—manifold morphological effects of single allelic difference	dead embryos or hatching eggs from heterozygous parents	7
Oryctolagus cuniculus (rabbit)	segregation: albino, light sable, dark sable—1:2:1	30s. per breeding pair	8a
Locusta migratoria	meiosis	all the year	11

E. Plants

i. *Genetics*

Organism	*Description and use*	*Availability (and season of use)*	*Source*
Antirrhinum majus	flower colours of defined genotype, and morphological mutants—segregation, linkage interaction	seeds (summer)	9a
Arabidopsis thaliana	wild types and mutants—segregation and mapping with short generation time	seeds (ephemeral)	12
Lycopersicum esculentum	normal, mutant and reciprocal translocation types—cytology and genetics of translocations	seeds (all seasons)	18
Lycopersicum, *Brassica*, *Matthiola*, *Rubus*, etc.	seedling characters	seeds	2b
Pisum sativum	20 mutants—segregation, pollination, emasculation	seeds (summer or in glasshouse)	1
Primula sinensis	five-point test cross	seeds	5
Rubus idaeus	*S*/*s* heterozygotes—segregation for glandular or eglandular in cotyledons but ratio distorted by semi-lethal linked to *s*	seeds (March/April)	17b

Organism	*Description and use*	*Availability (and season of use)*	*Source*
Senecio vulgaris	heterozygous at *ray* locus—1:2:1 segregation	seeds (not winter)	3
Solanum (potato) edible diploids	F2 seeds—segregation for two deleterious recessives showing as seedling characters	seeds (all seasons)	9b
Solanum (potato) edible diploids	pigment differences—by chromatography	tubers and chromatography schedule (Nov.–Mar.)	9b
Solanum demissum	seedling segregation for dominant hypersensitivity to virus Y	seeds and inoculation instructions (all seasons)	9b
Zea mays	Wx/wx—segregation in pollen	seeds (summer)	2e
Zea mays	genetic ratios in pericarp and endosperm characters	seeds	19
ii. *Chromosomes*			
Allium ursinum	diploid meiosis	fixed inflorescences	3
Cyanastrum johnstonii v. cuneifolium	partially asynaptic p.m.c.	Corms (Feb.–Mar.)	10
Endymion hispanicus	triploid meiosis	bulbs (February)	19
Haplopappus gracilis (2n = 4)	mitosis, meiosis	seeds (autumn)	2c
Lycopersicum esculentum	pachytene	seeds (summer)	18
Ornithogalum virens	(2n = 6)—root-tip mitosis and morphology of large chromosomes	bulbils (Jan.–Mar.)	10
Rumex hastatulus	(♂, $3^{II} + XY_1Y_2$, sex chromosomes	flowers (June)	2d
Secale cereale	inbred and interchange lines—structural and genetical variation in meiotic behaviour	seeds (June/July—can be fixed and stored)	15
Triticum aestivum	monosomics, trisomics, tetrasomics, mono- and di-isosomics, mono- and di-telosomics—aneuploidy	seeds (May/June flowering)	16

ii. *Chromosomes*			
Triticum aestivum and *Secale cereale*	lines with single pairs of *Secale* chromosomes added to *Triticum*—phenotypic effects of individual chromosomes	seeds (sowing autumn for summer demonstration)	16
Triticum Ægilops and *Secale*	natural variations of wild and cultivated spp. in these genera. Many synthetic allopolyploids	seeds	16
Tradescantia paludosa (2x)	meiosis, pollen mitoses	cuttings (May/Nov.)	2c, 3, 10
T. virginiana (4x)	meiosis, pollen mitoses	cuttings (May/Nov.)	2c
Tradescantia 3x (2x × 4x)	meiosis, pollen mitoses	cuttings (June)	2c
T. commelinoides (x = 8)	pollen germination in distilled water	cuttings at all times	10
iii. *Evolution*			
Agrostis tenuis	differences in lead tolerance between natural populations	tillers	3
Agrostis tenuis × *stolonifera*	demonstration of a successful sterile hybrid		
Brassica oleracea, *Beta maritima*, *Daucus carota*	characteristics of the wild progenitors of cultivated species	seeds	3
Festuca ovina	differences in alkaline/acid soil tolerance between natural populations	tillers	3
Solanum spp. and hybrids	natural and synthetic allopolyploids	tubers (plant in spring)	9b
iv. *Differentiation*			
Arabidopsis thaliana *Barbarea vulgaris*	variegation as Mendelian recessive nucleus acting on plastids	seeds (winter)	2e
Aubretia deltoides, *Thymus serpifolium*, *Salvia officinalis* (3 types)	periclinal chimaeras, due to gene or plastid mutation, showing occasional instability in new shoots	cuttings (all seasons)	2e
Delphinium ajacis	mutable gene	seeds (June/Sept.)	19

GENERAL SOURCES AND SERVICES

The following sources of materials are available apart from Research Laboratories:

BULBS, CORMS, ETC., of *Fritillaria*, *Crocus*, *Trillium*, and *Allium* spp. and in *Narcissus* both species and hybrids, notably the variety "Geranium", may be obtained from all bulb-growing firms. The names of species suitable for chromosome study can be discovered from the *Chromosome Atlas of Flowering Plants* (Darlington and Wylie 1956).

HORTICULTURAL PLANTS showing the principal types of variegation may be obtained from nurserymen dealing in ivy (*Hedera*), *Vinca*, *Salvia*, *Pelargonium*, and garden trees and shrubs and are listed in their catalogues.

PLANT SPECIES of botanical interest that are useful in genetic teaching may be obtained from Botanic Gardens through their seed lists or as material for propagation. These may be used for experimental breeding, like *Capsicum annuum*, for historical demonstration, like Mendel's *Mirabilis jalapa*, or for chromosome study, like *Rhoeo discolor* (ring of twelve), *Nicandra physaloides* (iso-chromosomes) or *Cestrum elegans* (heterochromatin and nucleic acid starvation) or *Allium* spp. (polyploidy and apomixis).

GRASSHOPPERS of the genus *Chorthippus* may be found during summer in grassland all over England. The cockroach, *Periplaneta americana*, studied by Lewis and John, may be obtained from all Biological Supply Agencies.

Those engaged on research may obtain information on other sources of material from the following international research services:

Drosophila Information Service
Wheat Information Service
Chromosome Information Service
The Human Chromosome News Letter
Microbial Genetics Bulletin
Mouse Letter

INDEX OF ADDRESSES

1. BARBER, Prof. H. N. Botany Dept., University, Hobart, Tasmania.
2. BOTANY SCHOOL, OXFORD
   (*a*) Dr. E. A. Bevan
   (*b*) Dr. L. K. Crowe
   (*c*) Prof. C. D. Darlington and C. G. Vosa
   (*d*) Dr. K. R. Lewis
   (*e*) Dr. R. A. E. Tilney-Bassett
3. BRADSHAW, Dr. A. D. Dept. of Agricultural Botany, Memorial Buildings, Bangor, Wales
4. DAWSON, MR. G. R., Dept. of Genetics, Trinity College, Dublin.
5. FROST, DR. L. C. Dept. of Botany, The University, Bristol 8.
6. HAMMERSMITH HOSPITAL, London, W.12. Director, MRC Microbial Genetics Research Unit.
7. HUNTON, P. Wye College, Ashford, Kent.
8. INSTITUTE OF ANIMAL GENETICS, Edinburgh 9.
   (*a*) Dr. R. A. Beatty
   (*b*) Dr. C. H. Clarke MRC Mutagenesis Unit.
9. JOHN INNES INSTITUTE, Bayfordbury, Hertford.
   (*a*) Dr R. J. S. Fincham Dept. of Genetics.
   (*b*) Dr. N. W. Simmonds Potato Genetics Dept.
   (*c*) B. Snoad Dept. of Applied Genetics.
10. JONES, DR. K. Royal Botanic Gardens, Kew, Surrey.
11. KEMP, R. F. O.* Dept. of Zoology, University College, Swansea.
12. MCKELVIE, A. D. North of Scotland College of Agriculture, Aberdeen.
13. MORGAN, DR. D. H. Dept. of Botany, The University, Hull.
14. REES, DR. H. Dept. of Agricultural Botany, University College of Wales, Aberystwyth.
15. RILEY, DR. R. Plant Breeding Institute, Trumpington, Cambridge.
16. ROPER, PROF. J. A. Dept. of Genetics, The University, Sheffield 10.
17. SCOTTISH HORTICULTURAL RESEARCH STATION, Invergowrie, Dundee.
   (*a*) Dr. G. Haskell Genetics Dept.
   (*b*) D. L. Jennings Genetics Dept.
18. SNEATH, DR. P. H. A. National Institute for Medical Research, London, N.7.
19. WHITEHOUSE, DR. H. L. K. Botany School, Downing Street, Cambridge.

* For Locusts also apply to the: Anti-locust Research Centre, 1 Princes Gate, London, S.W.7.

IV. CHROMOSOME FILMS

Two phase-contrast films showing chromosome movements in cell division are available.

1. *Mitosis in the Endosperm* (17 min.): film by A. Bajer of Cracow University, abridged, rearranged, and edited at the Botany School, Oxford.
2. *Meiosis in the Spermatocytes* (17 min.): two-reel film by Michel: to be hired from the Royal Microscopical Society, Tavistock House South, London, W.C.1.

These films may be hired for two days at 21*s*. plus postage.

V BOOKS ON THE TEACHING OF GENETICS

AUERBACH, C. 1953. *Notes for Introductory Courses in Genetics.* 3rd ed. Oliver & Boyd, London.

DEMEREC, M., AND KAUFMANN, B. P. 1957. *Drosophila. Guide for Introductory Studies of Genetics and Cytology.* 6th ed. Carnegie Inst., Washington.

GARDNER, E. J. 1952. *Genetics Laboratory Manual.* Burgess, Minneapolis.

HASKELL, G. 1961. *Practical Heredity with Drosophila.* Oliver & Boyd, Edinburgh.

HINTON, T. 1947. Suggestions for Laboratories in College Courses in Genetics. *Turtox News,* **25,** No. 1, 2.

MATHER, K. 1959. Genetics for Schools. John Murray: *Modern Science Memoirs*: No. 31.

ROYLE, H. A. 1960. *Laboratory Exercises in Genetics.* 4th ed. Burgess Publ., Minneapolis.

SHAW, G. W. 1960. John Murray: *Modern Science Memoirs:* No. 40.

STRICKBERGER, M. W. 1962. *Experiments in Genetics with Drosophila.* Wiley, New York & London.